Analog Circuits and Signal Processing

Series Editors:

Mohammed Ismail, Dublin, USA
Mohamad Sawan, Montreal, Canada

More information about this series at http://www.springer.com/series/7381

Rakesh Kumar Palani • Ramesh Harjani

Inverter-Based Circuit Design Techniques for Low Supply Voltages

 Springer

Rakesh Kumar Palani
Department of Electrical
  and Computer Engineering
University of Minnesota
Minneapolis, MN, USA

Ramesh Harjani
Department of Electrical
  and Computer Engineering
University of Minnesota
Minneapolis, MN, USA

ISSN 1872-082X   ISSN 2197-1854   (electronic)
Analog Circuits and Signal Processing
ISBN 978-3-319-83551-8   ISBN 978-3-319-46628-6   (eBook)
DOI 10.1007/978-3-319-46628-6

Printed on acid-free paper

This Springer imprint is published by Springer Nature
The registered company is Springer International Publishing AG
The registered company address is: Gewerbestrasse 11, 6330 Cham, Switzerland

Dedicated to Lord Muruga

Preface

Rapid advances in the field of integrated circuit design has been advantageous from the point of view of cost and miniaturization. Although technology scaling is advantageous to digital circuits in terms of increased speed and lower power, analog circuits strongly suffer from this trend. This is becoming a crucial bottle neck in the realization of a system on chip in scaled technology merging high-density digital parts with high-performance analog interfaces. This is because scaled technologies reduce the output impedance (gain) and supply voltage which limits the dynamic range (output swing). One way to mitigate the power supply restrictions is to move to current mode circuit design rather than voltage mode designs.

This thesis focuses on designing process, voltage, and temperature (PVT)-tolerant base band circuits at lower supply voltages and in lower technologies. Inverter amplifiers are known to have better transconductance efficiency, better noise, and linearity performance. But inverters are prone to PVT variations and have poor CMRR and PSRR. To circumvent the problem, we have proposed various biasing schemes for inverters like semi-constant current biasing, constant current biasing, and constant gm biasing. Each biasing technique has its own advantages, like semi-constant current biasing allows to select different PMOS and NMOS current. This feature allow for higher inherent inverter linearity. Similarly constant current and constant gm biasing allows for reduced PVT sensitivity. The inverter-based OTA achieves a measured THD of -90.6 dB, SNR of 78.7 dB, CMRR of 97 dB, and PSRR of 61 dB while operating from a nominal power of 0.9 V and at output swing of $0.9 V_{pp,diff}$ in TSMC 40 nm general purpose process. Further, the measured third harmonic distortion varies approximately by 11.5 dB with 120° variation in temperature and 9 dB with an 18 % variation in supply voltage.

The linearity can be increased by increasing the loop gain and bandwidth in a negative feedback circuit or by increasing the over drive voltage in open loop architectures. However both these techniques increases the noise contribution of the circuit. There exist a trade off between noise and linearity in analog circuits. To circumvent this problem, we have introduced nonlinear cancellation techniques and noise filtering techniques. An analog-to-digital converter (ADC) driver which is capable of amplifying the continuous time signal with a gain of 8 and sample onto the input capacitor (1 pF) of 1 10 bit successive approximation register (SAR) ADC is designed in TSMC 65 nm general purpose process. This exploits the non-linearity cancellation in current mirror and also allows for higher bandwidth operation by decoupling closed loop gain from the negative feedback loop. The noise from the out of band is filtered before sampling leading to low noise operation. The measured design operates at 100 MS/s and has an OIP_3 of 40 dBm at the Nyquist rate, noise power spectral density of $17\,nV/\sqrt{Hz}$, and inter-modulation distortion of 65 dB. The intermodulation distortion variation across ten chips is 6 and 4 dB across a temperature variation of 120 °C.

Non-linearity cancellation is exploited in designing two filters, an anti-alias filter and a continuously tunable channel select filter. Traditional active RC filters are based on cascade of integrators. These create multiple low impedance nodes in the circuit which results in a higher noise. We propose a real low pass filter-based filter architecture rather than the traditional integrator-based approach. Further, the entire filtering operation takes place in current domain to circumvent the power supply limitations. This also facilitates the use of tunable non-linear metal oxide semiconductor capacitor (MOSCAP) as filter capacitors. We introduce techniques of self-compensation to use the filter resistor and capacitor as compensation capacitor for lower power. The anti-alias filter designed for 50 MHz bandwidth that is fabricated in IBM 65 nm process achieves an IIP3 of 33 dBm while consuming 1.56 mW from 1.2 V supply. The channel select filter is tunable from 34 to 314 MHz and is fabricated in TSMC 65 nm general purpose process. This filter achieves an OIP3 of 25.24 dBm at the maximum frequency while drawing 4.2 mA from 1.1 V supply. The measured intermodulation distortion varies by 5 dB across 120 °C variation in temperature and 6.5 dB across a 200 mV variation in power supply. Further, this filter presents a high impedance node at the input and a low impedance node at the output easing system integration.

SAR ADCs are becoming popular at lower technologies as they are based on device switching rather than amplifying circuits. But recent SAR ADCs that have good energy efficiency have had relatively large input capacitance increasing the driver power. We present a 2X time interleaved (TI) SAR ADC which has the lowest input capacitance of 133 fF in literature. The sampling capacitor is separated from the capacitive digital to analog converter (DAC) array by performing the input and DAC reference subtraction in the current domain rather than as done traditionally in charge domain. The proposed ADC is fabricated in TSMC's 65 nm general purpose process and occupies an area of $0.0338\,mm^2$. The measured ADC spurious free dynamic range (SFDR) is 57 dB, and the measured effective number of bits (ENOB) at Nyquist rate is 7.55 bit while using 1.55 mW power from 1 V supply.

Minneapolis, MN, USA   Rakesh Kumar Palani
                                                                                         Ramesh Harjani

Acknowledgments

There are many people who have helped, supported, and guided me to reach this juncture in my life, and this thesis will be incomplete without offering my gratitude to them. Firstly, I would like to express my gratitude to Prof. Ramesh Harjani for giving me the opportunity to work in his lab and guiding me through my PhD. I am extremely thankful for the freedom he provided me to pursue the research way I liked. But nevertheless, he always knew what I did and advised me when he felt I was spending too much time on something unreasonable or chasing something unachievable. From my conversations with my friends in other universities, I have realized that I have been very lucky in terms of number of tape-outs and the testing facilities at my disposal. I am indebted to Prof. Harjani for making these facilities available to us and not have to worry about these things. I have not only benefited technically but also learned many other skills such as technical writing, making quality presentations, and proposal writing. I would like to thank Savita Ma'am for her wonderful lecture on getting things done.

I would like to thank Prof. Chris Kim, Prof. Anand Gopinath, and Prof. Hubert Lim for agreeing to chair my defense and preliminary oral committee. The comments given by the committee during the preliminary oral exam were very insightful and helpful toward the latter half of my PhD.

I am thankful to Martin for the discussions we had and also for his help in teaching me soldering and the use of lab equipment. I am thankful to Martin for teaching me paper writing and presentation skills along with his grammar correction.

I am grateful to them for sharing the intense workload during the tape-outs. I am also thankful to Sriharsha Vadlamani for giving me a coffee company throughout my PhD. Many ideas in my PhD is evolved from the coffee discussions.

I am grateful to DARPA for supporting my project. I am thankful to the DARPA CLASIC Team, Prof. Danijela, Prof. Dejan Markovic, and Fang Li Yuan, for helping me in understanding the digital back end of a software-defined radio.

I learned a lot during my internships at Broadcom and Qualcomm. I was able to imbibe some of the design methodology and practices that ensure the success of design in my project as well. I am indebted to my managers Myron Buer,

Carl Monzel, and Yifei zang at Broadcom for teaching me the memory design and testing with focus on PVT variations. I am also thankful for my mentor Wang Yan at Qualcomm who introduced me to discrete time delta sigma modulators and also Ganesh Kiran , Liang Dai, and Dinesh Alladi for technical discussions.

There are many people in the ECE Department who silently work behind the scenes and ensure that we students always have the best possible facilities. They ensure that we spend minimum time on the administrative work and focus most of the time on our research. I would like to thank Carlos Soria, Chimai Nguyen, Dan Dobrick, Becky Colberg, Linda Jagerson, Kyle Dukart, Jim Aufderhar, Paula Beck, and Linda Bullis.

Finally, I am ever grateful to my parents for believing in me and supporting me in every endeavor that I undertook. They always ensured the best for me and gave me all the freedom to do what I liked the most. My brother has been funny and a good companion for the last three years of my PhD.

I sincerely thank you all!

Contents

List of Figures

List of Tables

Chapter 1
Introduction

The increase in chip complexity over past few years has created the need to implement complete analog and digital subsystems on the same integrated circuit using the same technology. Figure 1.1 shows the roadmap for the technology scaling. The increase in demand for battery operated portable devices and implantable medical devices has placed added pressure on lowered supply voltages. Technology scaling reduces the delay of the circuit elements, enhancing the operating frequency of an integrated circuit. The density and number of transistors on an IC increases with the scaling of the feature sizes. Today we are at 14 nm FINFET technology. Reducing power dissipation has become an important objective in the design of digital circuits. One common technique for reducing power is to reduce the supply voltage. Reduction in supply voltage demands proportional scaling of threshold voltage to maintain the same ON current. However scaling of threshold voltage increases the sub threshold leakage or the OFF current. Hence threshold voltage does not scale proportional to the supply voltage. Technology scaling (Fig. 1.1) is a robust roadmap (www.itrs.net.) for digital circuits, while analog circuits strongly suffer from this trend, and this is becoming a crucial bottle neck in the realization of a system on chip in a scaled technology merging high-density digital parts, with high performance analog interfaces. This is because scaled technologies reduce the supply voltage, and this limits the analog performance in qualitative (is it possible to operate from a low voltage?) and quantitative (if it is possible to operate, which performance is achievable?) terms [1].

Portable devices like mobiles (Fig. 1.2) continue to drive the need for circuits that achieve low power without sacrificing linearity. Analog baseband circuits, including filters and programmable gain amplifiers (PGA), are indispensable in wireless sensors and communication systems. These analog filters typically consume tens of mWs of power and have a considerable impact on the total power consumption. Hence implementation of analog functions in MOS technology has become increasingly important, and great strides have been made in implementing functions such

© Springer International Publishing AG 2017
R.K. Palani, R. Harjani, *Inverter-Based Circuit Design Techniques for Low Supply Voltages*, Analog Circuits and Signal Processing,
DOI 10.1007/978-3-319-46628-6_1

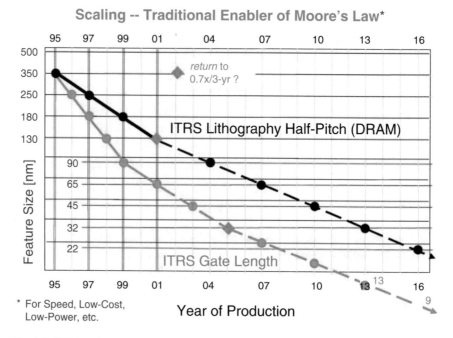

Fig. 1.1 ITRS roadmap

Fig. 1.2 Development in mobile industry

as ADCs, DACs, filters, voltage references, instrumentation amplifiers in CMOS technology. Operational transconductance amplifiers (OTAs) are widely employed as active elements in filters, data converters and buffer amplifiers.

Each mobile will have many radios. One typical simplest RF receiver chain is shown in Fig. 1.3. These has couple of filters for filtering different band signals and also has amplifiers to amplify inband signals. The small signal from antenna is band selected using an off chip passive band pass filter. Low noise amplifier (LNA) is used to amplify these signals with minimal noise addition. The signal at RF carrier

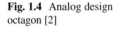
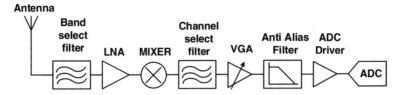

Fig. 1.3 Typical RF receiver

Fig. 1.4 Analog design
octagon [2]

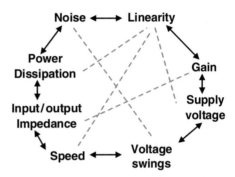

frequency is down converted to baseband using a mixer. A channel select filter is used to select the signal channel. A cascade of IF amplifiers and VGA is used to amplify the signal. Anti alias filter is used to remove all the components away from the Nyquist band. This is used to prevent aliasing of out of band signals and noise into the signal band after sampling. If the signal swing is less than the dynamic range of ADC (rail-rail), an ADC driver can be used to amplify and sample the signal onto the input capacitance of the ADC.

The trade of in analog circuit design is explained in Fig. 1.4 [2]. The parameters like gain, speed, power dissipation, supply voltage, linearity, noise and maximum voltage swings are important in analog design and trades off with each other. Furthermore, the input and output impedances determine how the circuit interacts with the preceding and subsequent stages. For example at lower supply voltages, we are hit by the noise floor. Hence we need to operate the transistors at lower overdrive voltages for better noise performance which in turn hurts the linearity of the transistor. Similarly we need to burn more power to reduce the noise and increase the speed of amplifier. The gain, supply voltage and impedances along with voltage swings determine the maximum signal to noise ratio achievable from the circuit. Similarly higher linearity demands higher overdrive voltage which increases the noise contribution. This book focuses on the design of baseband circuits in a wireless receiver like amplifiers, channel select filter, anti alias filter and time interleaved ADC. The circuits are optimised for lower noise and techniques like non linear cancellation are used to increase the inherent linearity. Further filter circuits are designed in current mode where both low noise and higher linearity demands higher overdrive voltages. The design of high performance baseband circuits in MHz range is always challenging. It is difficult to get the negative feedback loop gain at these frequencies due to higher threshold voltage at lower power supply and also with lower output impedances. This book provides a different architecture for filters to achieve high linearity and low noise at lower power. Further the channel select filters is made tunable to select the channels from 34 MHz to as wide as 314 MHz. The ADC driver is designed with a gain of 8 to increase the swing of the signals to rail to rail and sample onto the input capacitance of the ADC. Finally a time interleaved ADC is designed to convert analog to digital for signal processing. This ADC offers high impedance to the preceding circuits and thereby lowering the power of the entire system.

1.1   Traditional Operational Transconductance

Operational amplifier is required to realize an integrator in negative feedback circuit. Since the loop gain of the negative feedback circuit determines the performance of the circuit, design of operational amplifier is an hot area of research in analog VLSI circuits. In fully differential circuits, the operational amplifiers suppresses common mode differences. The simplest operational amplifier is a five transistor differential pair (Fig. 1.5). This forms the core in more complex operational amplifier design.

   We apply the input voltage across the differential pair transistors. The tail transistor (biased at ntail) acts like a current source thereby acting like a source degeneration resistor for common mode signals. Hence the differential pair transistors convert only the differential mode components in input voltage to current. The common mode voltage appears directly at the tail node. This current is pumped into output impedance of the transistors through a current mirror to get voltage gain. The finite output impedance of the transistors limits the gain of the circuit. Hence attempts were made to improve the output impedance of the transistors. A common gate amplifier has the low input impedance due to inherent negative feedback but higher output impedance. Hence the current from the common source differential pair acts like an input to common gate amplifier. This leads to an architecture

Fig. 1.5 Five transistor differential pair

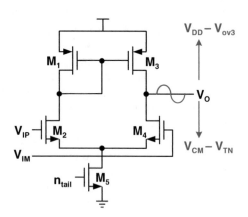

Fig. 1.6 Telescopic folded
OTA

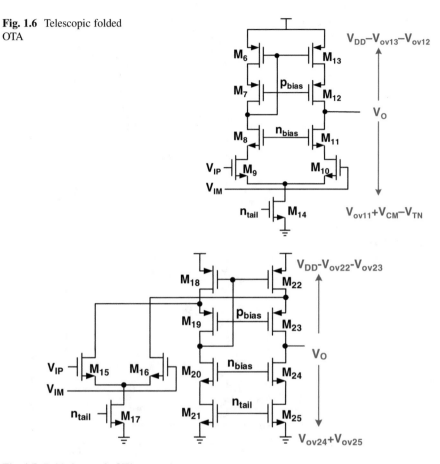

Fig. 1.7 Folded cascode OTA

of telescopic cascoded operational amplifier (Fig. 1.6). The output impedance of the OTA is amplified by the gain of the common gate amplifier. This leads to have an increased gain. The output swing is small as each transistor requires overdrive voltage to maintain them in saturation region. Folded cascoded amplifier is introduced to increase the output swing by one overdrive voltage. Here instead of cascading NMOS based common source amplifier with NMOS based common gate amplifier, we cascade NMOS common source amplifier with PMOS common gate amplifier (Fig. 1.7). This architecture gives increased gain but at the cost of increased power and noise. The advantage of this architecture is decoupled input and output common mode voltages and increased output swing.

The cascode architectures take current converted by the input differential pair transistors through a low impedance nodes. Hence these typically do not require compensation as the poles created at the low impedance nodes are at higher frequencies. A current mirror based OTA is proposed (Fig. 1.8), to increase the gain

Fig. 1.8 Current mirror OTA

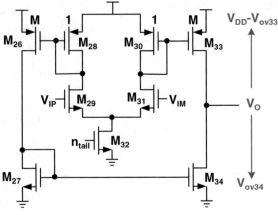

Fig. 1.9 Two stage
telescopic cascoded OTA

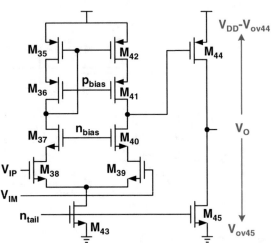

with the aid of number of fingers in current mirror. Here the gain is increased by
increasing the number of fingers in the current mirror. This in turn increases the bias
current which result in higher power consumption.

The common source telescopic amplifier can be cascaded with another common
source amplifier to obtain higher gain (Fig. 1.9). We obtain the maximum possible
swing in this architecture as the second stage common source amplifier has only one
PMOS and NMOS transistor. The swing at the output of telescopic cascoded stage
is reduced by the gain of the second stage. This amplifier gives highest gain, lowest
noise. The internal high impedance node at the cascade point of common source
amplifiers results in a low frequency pole which has to be compensated.

Table 1.1 shows the comparison between various operational amplifier architectures. H represents high and L represents Low in this table. If the linearity of the
operational amplifier is dominated by the input differential pair, all the architectures
has similar linearity performance. Similarly we assume all the OTA architectures

Table 1.1 Comparison of traditional OTA architectures

	Five transistor	Telescopic	Folded cascode	Current mirror	Two stage
Output swing	H	L	HH	HHH	HHH
Freq. response	HHH	HH	H	HH	L
Gain	L	HH	HH	H	HHH
Noise	L	L	HHH	H	L

are designed in same technology with identical power supply. The output swing is highest in current mirror OTA and two stage design. Two stage design gives the highest possible output swing, gain and contributes lower noise but suffers from the low frequency response. Hence this architecture is preferred mostly in high resolution negative feedback loops.

The scaling of power supply makes the design of differential pair difficult. Further the poor output impedance results in poor CMRR and PSRR. Since differential pair becomes the core of any OTA architecture, it is compared with the proposed inverter based OTA in Sect. 1.5. A two stage inverter based two stage telescopic operational amplifier is proposed for high resolution applications.

1.2 Differential Pair Versus Inverter

In conventional differential pair based OTAs, the minimum input common mode voltage is bounded by a threshold voltage and the overdrive voltages of the differential pair plus that of the tail source limiting the input voltage swing. Input and output common mode voltages are equal in a typical continuous time systems to avoid any common mode currents in the system. Hence the input common mode limitation restricts the overall output swing in the system. Lower supply voltage severely constraints the tail current overdrive voltage deteriorating CMRR and also prevents the use of cascode devices limiting gain. Further, the large signal linearity of differential pairs is limited by the finite tail current. Body input OTA designs is proposed in [3, 4] at lower supply voltages. However this results in lower frequency response and also increased non-linearity. Current reuse in inverters enables at least a 2X higher transconductor (g_m/I_d) efficiency compared to a differential pair. Inverters allow rail-to-rail input swing because of the class AB operation. Hence the input and output common mode can be at mid supply for optimal signal swing at lower supply voltages. The poor PVT tolerance, CMRR and PSRR challenges inverter based designs. Non-cascoded inverter based OTA designs with common mode cancellation was proposed in [5, 6]. Linearity improvement using cancellation cancelation techniques have been proposed for higher supply voltages [7–10]. Use of ring oscillators as amplifiers in switched capacitor circuits is proposed in [11].

Figure 1.10 shows a comparison of a traditional differential pair and a pseudo differential inverter. The bias current in both the designs is assumed to be equal

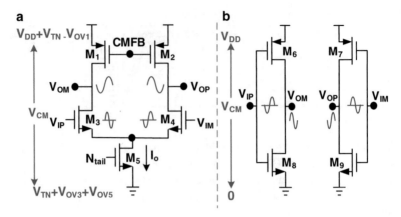

Fig. 1.10 Input and output swings of (**a**) differential pair and (**b**) inverter OTAs

to I_o. The minimum input common mode voltage for the differential pair is given by
Eq. (1.1) and optimal common mode voltage for inverter is half the power supply.

$$V_{CM,diff,min} = V_{T3} + V_{ov3} + V_{ov5} \tag{1.1}$$

The threshold voltage of the transistor $M_{3,4}$ is higher than that of M_{6-9} due to the
body effect. This results in lower swing as described by a simulation example in
TSMC's 65 nm CMOS technology. We will use some typical numerical numbers to
illustrate our example. The overdrive voltage (V_{ov}) of all the devices are assumed
to 125 mV. Due to body effect the transistors (M3, M4) have a threshold voltage
(V_{TN}) of 440 mV (50 mV above nominal). Therefore, the minimum input common
mode voltage for the differential pair is $V_{T3} + V_{OV3} + V_{OV5} = 690$ mV. This clearly
limits the input signal range. Furthermore, in most continuous time systems, that are
once again becoming popular due to the limited headroom for switches, the input
and output common mode voltages become equal due to the DC negative feedback
around the loop. Hence the minimum output common mode is also 690 mV. With a
power supply of 0.9 V and one overdrive drop at the PMOS transistor, the maximum
attainable swing is now 170 mVpp. Inverter based designs ($M_{6,8}$ and $M_{7,9}$) allow rail-
to-rail input swing because of the class AB operation. This translates to a maximum
attainable output swing of 650 mVpp (4x larger than traditional OTAs). Further the
transistors do not suffer from body effect resulting in higher linearity and trans
conductance.

1.3 Non Linearity Analysis

The linearity of a trans conductor is limited by its trans conductance linearity and output impedance linearity. The differential output current in the differential pair can be derived by assuming square law model for the transistors M_{3-4} as Eq. (1.2)

$$I_{out} = -(V_{IP} - V_{IM})\beta_n \sqrt{\frac{2I_o}{\beta_n} - (V_{IP} - V_{IM})^2} \qquad (1.2)$$

where $\beta_n = \mu_n C_{ox} W/2L$, μ_n is the mobility of the electrons, C_{ox} is the oxide capacitance and W and L are the width and length of the devices. For smaller input voltages $(V_{IP} - V_{IM} \approx 0)$, the output current is given by

$$I_{out,diff} = -G_{m,diff}(V_{IP} - V_{IM}) \qquad (1.3)$$

where $G_{m,diff} = \sqrt{2\beta_n I_o}$. Equation (1.2) also suggests that G_m falls to zero for $(V_{IP} - V_{IM}) = \sqrt{2I_o/\beta_n}$. The output current has odd order harmonics and even order harmonics are suppressed by the differential operation. The odd order harmonics created is a result of current limitation with tail current source M_5 in differential pair. Although the tail current source biases the differential pair at constant current and also give common mode rejection ratio, this results in non linearity. Further the body effect in transistors $M_{3,4}$ also increases the non-linearity.

However the pseudo-differential output current of inverter based amplifier is given by Eq. (1.4)

$$I_{out,inv} = -(V_{IP} - V_{IM})(\beta_p [V_{DD} - V_{CM} - V_{TP}] + \beta_n [V_{CM} - V_{TN}]) \qquad (1.4)$$

The output differential current for an inverter with transistors obeying square law is highly linear as all the even order harmonics are suppressed by the differential operation. The small signal trans conductance is given by

$$G_{m,inv} = (\beta_p [V_{DD} - V_{CM} - V_{TP}] + \beta_n [V_{CM} - V_{TN}]) \qquad (1.5)$$

Figure 1.11 shows the output current of the differential pair and pseudo differential inverter with identical small signal transconductance. The tail current in the differential pair saturates the current to I_o resulting in nonlinearity. However the output current in an inverter increases with the input voltage due to its class AB operation. The non linearity in the output current of the inverter is primarily due to its short channel effects and its deviation from square law model.

In analog design the channel length is typically selected to be higher than the minimum to increase the output impedance of the transistor. The transistors in output stage of OTA design typically has a smaller channel length to reduce the parasitic capacitance and also to create a low impedance output node. The gain of

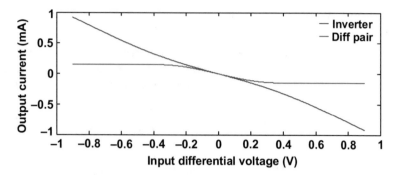

Fig. 1.11 Output current of a differential pair and pseudo-differential inverter

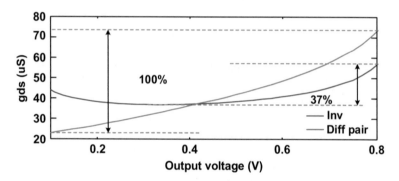

Fig. 1.12 Output impedance variation with output swing in differential pair and inverter

this stage is typically between 5 and 10. For amplifiers driving larger load currents, the non-linearity in the output impedance becomes significant. Figure 1.12 shows the output impedance variation with output swing for class A (differential pair) and class AB (inverter) amplifiers. The output impedance of a transistor decreases with increase in the current. Hence for differential pair the conductance increases with the swing. However for inverter the PMOS current increases and NMOS current decreases with output swing resulting in lower output impedance variation. Although the output current in inverter based amplifiers are linear, unlike differential pair it strongly depends on the input common mode voltage which restricts the use of inverter based designs.

1.4 Noise Analysis

The input referred noise for a differential pair and for a pseudo differential inverter is given by Eq. (1.6).

$$v_{n,diffpair}^2 = \frac{8kT\gamma}{gm_3}\left(1 + \frac{gm_1}{gm_3}\right) \tag{1.6}$$

$$v_{n,inv}^2 = \frac{8kT\gamma}{gm_6 + gm_8} \tag{1.7}$$

The transconductance gm_3 is assumed to be equal to the inverter transconductance $gm_6 + gm_8$ for the sake of comparison. The excess noise factor for the inverter is 1 which is less than that for the corresponding differential pair $[(1 + gm_1/gm_3)]$. This is because all the transistors in the inverter contribute both to the signal and to the noise whereas in the differential pair the load transistor (M_1 and M_2) contribute only to the noise.

A doubling in the width of both the PMOS and NMOS transistors does not change its gain. It is equivalent to adding the gm cells in parallel where both gm and gds increases by same amount. Hence only the channel length determines the gain of the inverter. Any increase in the width of the transistor results in an increase in its *gm* resulting in an increase in the system UGF. This property of inverter based designs separates the gain and *gm* parameters simplifying design. Simulations show that with constant gm biasing, the effective gds varies less than 20 % across PVT variations.

Inverter based amplifier supports higher signal swings with higher linearity and lower noise compared to differential pair based amplifiers. This makes the inverter amplifiers attractive especially at lower technologies and lower power supplies. However the dependence of the inverter amplifier's bias voltage and currents with PVT restricts their use in modern technologies.

1.5 Inverter Transconductor

Figure 1.13 shows the inverter transconductor circuit from Nauta [6, 12]. The inverters $Inv_{1,3,5}$ are identical to those of the differential counterpart $Inv_{2,4,6}$. The common mode level of the output voltages V_{OP} and V_{OM} is controlled by the four inverters Inv_{3-6}. The output common mode voltage is at the meta stable point of

Fig. 1.13 Nauta inverter transconductor

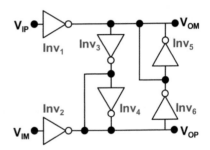

Fig. 1.14 Inverter based 2
stage OTA

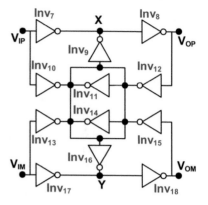

the inverters ($Inv_{4,5}$). The common mode and differential mode impedance offered
by these inverters are $1/(g_{m3} + g_{m4})$ and $1/(g_{m4} - g_{m3})$. The common mode gain is
given by Eq. (1.8) and the differential mode gain is given by Eq. (1.9)

$$A_{cm} \approx \frac{g_{m1}}{g_{m3} + g_{m4}} \tag{1.8}$$

$$A_d = \frac{g_{m1}}{(g_{m4} - g_{m3}) + g_{ds1} + g_{ds5} + g_{ds6}} \tag{1.9}$$

Inverters (Inv_{3-6}) are designed to offer negative impedance to differential signals
by making g_{m3} greater than g_{m4}. This is used to increase the differential mode
gain by increasing the effective differential impedance. The transconductance has
a large bandwidth because of the absence of internal nodes [6]. The inverter
transconductance in this design is set by altering the supply voltage and hence
requires an on chip power regulator. Tunable inverters using body terminal control in
a master slave approach was proposed in [13]. A two stage inverter based differential
OTA is shown in Fig. 1.14 [5]. The first stage has feedforward paths (Inv_{9-11}) for
common mode cancellation, while the second stage uses additional feedback paths
for the common-mode ($Inv_{9,11,12}$). The transconductances of inverters (Inv_{7-12}) are
identical to those of inverters (Inv_{13-18}) for fully differential operation. The input
common mode voltage (ΔV_{cm}) generates a current of $(g_{m7} - g_{m9}g_{m10}/g_{m11})\Delta V_{cm}$
at node X and Y. If the transconductance ($g_{m9}g_{m10}/g_{m11}$) is made equal to g_{m7} as
in Eq. (1.10), then the voltages at node X and node Y are invariant to any input
common mode variations.

$$g_{m7} = \frac{g_{m9} \cdot g_{m10}}{g_{m11}} \tag{1.10}$$

However, unlike a traditional differential pair where only the differential mode
components are converted to current, here both the differential and common mode
components are converted to current and only at the outputs are the common

mode currents are cancelled. The common mode transfer function from node X to output V_{OP} is given by Eq. (1.11).

$$\frac{V_{OP}}{X} = \frac{g_{m8}/g_{ds8}}{1 + (g_{m8}g_{m9}g_{m12})/(g_{ds9}g_{m11}g_{ds8})}$$

$$\approx \left(\frac{g_{m11}}{g_{m12}}\right)\frac{g_{ds9}}{g_{m9}} \tag{1.11}$$

The common mode signals are suppressed in the feedforward path and then further suppressed by the gain given by Eq. (1.11) in the feedback path. Unlike the Nauta transconductor, this design does not increase the output impedance using negative resistance but instead uses higher impedance nodes. Hence, the differential mode gain is given by Eq. (1.12).

$$A_d = \frac{g_{m7}g_{m8}}{(g_{ds7} + g_{ds9})g_{ds8}} \tag{1.12}$$

Further this design requires frequency compensation for both the differential and common mode feedback loops. Circuit simulations were used to show a low frequency CMRR of 65.8 dB at 1.8 V along with a differential mode gain of 48.2 dB in [5]. Since the metastable point of the inverter varies with PVT, the designs in [5, 6, 12, 13] and other inverter based designs [14–16] are sensitive to PVT variations.

1.6 Non-linearity Cancellation Techniques

If a function is multiplied by its inverse function, then any non-linearity in the function is canceled (i.e. $ff^{-1} = 1$). This property is widely exploited to cancel the nonlinearity of the transistors. Figure 1.15 shows different circuit techniques which uses inverse function to cancel the nonlinearity in transistors. Figure 1.15a is a common source NMOS amplifier with a NMOS load. Transistor (M_2) converts voltage to current with transconductance g_{m2} and load M_1 converts this current to voltage using resistance $1/g_{m1}$. The resistance $1/g_{m1}$ is an scaled inverse function of transconductance g_{m2} as they are from identical scaled NMOS devices. This makes the output voltage linear with input voltage irrespective of any transconductance nonlinearity in transistors. Figure 1.15b is a common source NMOS amplifier with diode connected PMOS load to reduce any body effect. Since both PMOS and NMOS transistors are square law devices the transconductance nonlinearity of NMOS M_4 can be partly cancelled by sizing PMOS transistor M_3 appropriately. Figure 1.15c shows a NMOS based current mirror exploiting nonlinear cancellation. M_5 converts current to voltage based on its inverse non linear function f_1^{-1} and M_6 converts this voltage back to current by using the same function f_1 making the output current to be linear with input current.

Fig. 1.15 Traditional
non-linearity cancellation
techniques

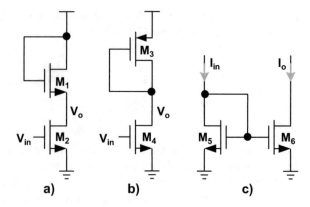

This principle is used in the design of filters [17] and ADC driver [18] circuits. The nonlinear transconductance cancellation technique can also be used to increase the inherent linearity of inverters. Using square law model, the output current is given as the difference between the NMOS and PMOS current as

$$I_o = (\beta_n - \beta_p)V_{in}^2 - 2V_{in}\left(\beta_n V_{TN} - (V_{DD} - V_{TP})\beta_p\right)$$
$$+ V_{TN}^2 \beta_n - V_{TP}^2 \beta_p \tag{1.13}$$

The square nonlinearity can be canceled by selecting the β factor of NMOS and PMOS transistor equal. The fully differential implementation inherently cancels the even order harmonics leading to highly linear operation. When the input signal swing is large, one transistor enters subthreshold regime. The nonlinearity for short channel devices can be cancelled by the choice of bias current in the transistors.

1.7 Organization

The main focus of this book is to develop inverter based baseband circuits which is tolerant to PVT variance. The circuits are required to be highly linear, fully integrated on chip with low noise performance and low power consumption. Further all the circuits should be easily system integrable, like it should have high input impedance and low output impedances. Most of the circuits developed in this book is self compensated, requiring no additional compensation capacitor. Cascoding of inverter amplifier and use of external common mode feedback circuitry to get CMRR and PSRR are verified by designing OTA in TSMC's 40 nm GP process. The circuit techniques like self compensation, non linear cancellation are verified by designing ADC driver, filter and SAR ADC in TSMC's 65 nm CMOS process.

Chapter 2 focuses on the biasing technique for inverter amplifiers. Here we start with the basic metastable biasing of the inverter and then discuss about the short coming of this type of biasing. We then propose semi constant current biasing of inverter where only the NMOS transistor in inverter is biased at constant current. This reduces the PVT sensitivity and also enables the use of inverter with different PMOS and NMOS currents. Hence the NMOS current can be selected to increase the linearity of the overall inverter trans conductance. Then we introduce a constant current biasing and constant gm biasing of inverter. Simulations were performed in TSMC's 65 nm GP process to verify the PVT sensitivity of constant current and gm biasing of inverter.

Chapter 3 discusses about the design of all inverter based fully differential operational amplifier design. We introduce an external negative feedback to suppress the common mode signals at the input. This gives a good CMRR and PSRR for the OTA. We then discuss about cascoding an inverter and technique of current reference free cascode biasing. he OTA is fabricated in TSMC's 40 nm GP process with 0.9 V supply, achieves a THD of 90.6 dB, SNR of 74 dB, CMRR 97 dB and PSRR 61 dB over 10 MHz while driving a 2 pF load. The common mode feedback biases the first and second stage inverters of OTA at their metastable voltage. This maintains the phase margin and stability over the temperature and power supply variations. We get a variation in measured third harmonic distortion at 9.5 MHz to be around 11.5 dB with 120° variation in temperature and 9 dB across 18 % variation in power supply.

Chapter 4 introduces to the design of current mirror based circuits. We designed a ADC driver to sample rail to rail input onto 1 pF capacitor for a 10 bit 100 MS/s ADC. We exploit the non linear cancellation of a current mirror. Further we introduce techniques like sampling through series resistor to filter out of band noise and also to compensate the negative feedback loop. As a proof of concept an ADC driver is designed and implemented in TSMC's 65 nm GP CMOS technology. The measured design operates at 100 MS/s and has an OIP_3 of 40 dBm at the Nyquist rate, provides a gain of 8, and samples the signal onto a 1 pF output capacitance while drawing 2 mA from a 1 V supply.

Chapter 5 describes a third order Butterworth anti-alias filter design based on real low pass filter architecture rather than traditional integrator based approach. This reduces the power consumption and noise contribution by reducing the number of low impedance nodes in the circuit. This filter is in current mode and exploits the non linear cancellation of current mirror for obtaining linear gain. We introduce the concept of self compensation in filters. The load acts as the compensation capacitance to the OTAs allowing the majority of the current to flow into the load, increasing the overall power efficiency. As a proof of concept a third order filter is fabricated in IBM 65 nm technology. The measured prototype designed for a 50 MHz bandwidth achieves an IIP3 of +33 dBm and 1.8X better FOM over state-of-art while drawing 1.3 mA from a 1.2 V supply, is capable of driving a 1 pF load, and occupies 6X smaller area.

Chapter 6 further extends the concepts of Chap. 5 to build a continuously tunable channel select filter. We only use non linear MOSCAPs as filter capacitance in this design. As a proof of concept 30 −314 MHz tunable filter is fabricated in TSMC's 65 nm GP process. Although filter uses MOSCAPs, it achieves IIP_3 of

22 dBm at the highest tuning frequency. Further all the negative feedback circuits are self compensated using the filter resistor and capacitor resulting in low power of 4.6 mW and 17.5x smaller area. Due to the biasing of inverter with semi constant current biasing and owing to modular design, IMD varies only by 6.5 dB over 200 mV variation in power supply and 5 dB across temperature. The filter achieves the highest figure of merit among the state of art published filters.

Chapter 7 focuses the design of 220 MS/s time interleaved SAR ADC. We exploit the non linearity cancellation between the input and the reference path in preamp for achieving high linearity. The proposed ADC fabricated in TSMC's 65 nm GP process occupies an area of $0.0338\,mm^2$ and consists of two time-interleaved channels each operating at 110 MS/s. The sampling capacitor is separated from the capacitive DAC array by performing the input and DAC reference subtraction in the current domain rather than as done traditionally in the charge domain. This allows for an extremely small input capacitance of 133 fF. The measured ADC SFDR is 57 dB and the measured ENOB is 7.55 bits at Nyquist rate while using 1.55 mW power from 1 V supply.

Chapter 2
Biasing

Inverter amplifiers have traditionally been biased using a constant voltage replica biasing technique. The replica is typically an equal sized inverter with input and output shorted [6]. This method of biasing ensures that the inverters are biased at their maximum transconductance (g_m) and that the input and output common mode voltages remain equal at V_M. Unfortunately, this method of replica biasing has its limitations as the bias point is directly affected by PVT variations. In fact, using this technique the effective transconductance can vary $\approx$ 40 % with PVT variations impacting bandwidth, stability and gain. To solve this problem we introduce three techniques, semi-constant current biasing (SCCB), constant current biasing and constant gm biasing [19]. We have verified the semi-constant current biasing technique using multiple fabricated designs, while the constant current and gm biasing technique have only been verified using circuit simulations. Therefore, we first introduce the semi-constant current biasing technique and evaluate its performance. This is followed by constant current and gm biasing techniques.

2.1 Semi-constant Current Biasing

To implement SCCB the inverters are skewed in size such that even at the PMOS fast - NMOS slow corner, the transconductance for the NMOS is greater than for the PMOS. We need this additional degree of freedom to control PVT variations. In our design the NMOS transistor is the same size as the PMOS transistor. This choice gives up some transconductance efficiency ($\approx$30 % for $\mu_n/\mu_p = 2.5$) for increased PVT tolerance. The NMOS in the main unit inverter (IU) is biased with a constant current as shown in Fig. 2.1 [18]. The W/L size of the NMOS transistor is selected such that the gate voltage (V_b) is close to mid supply. For the nominal supply voltage the voltage V_m is equal to V_b due to the OTA$_b$ feedback loop. The auxiliary inverter

© Springer International Publishing AG 2017
R.K. Palani, R. Harjani, *Inverter-Based Circuit Design Techniques for Low Supply Voltages*, Analog Circuits and Signal Processing,
DOI 10.1007/978-3-319-46628-6_2

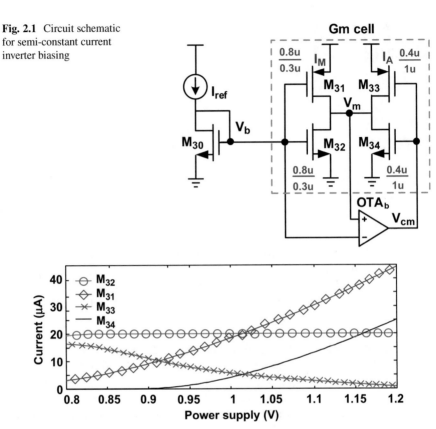

Fig. 2.1 Circuit schematic for semi-constant current inverter biasing

Fig. 2.2 Biasing network current with power supply variation

(I_A) is used to make the input and output voltage of the unit inverter equal using negative feedback. This is necessary to ensure that the main inverter (I_M) remains in saturation and it also makes the cascading of inverters possible. Further, this reduces any drain source voltage mismatch between the NMOS transistor in the main inverter (M_{32}) and the diode connected NMOS (M_{30}). Figure 2.2 shows the current in the different transistors with changes in the power supply. At the nominal supply voltage the current in the NMOS (M_{32}) is higher than in the PMOS (M_{31}) making the NMOS transconductance higher than for the PMOS, recall that both devices are sized the same. The NMOS (M_{32}) in the main inverter is biased at a constant current and hence it is constant with power supply. An increase in the supply voltage increases the PMOS (M_{31}) current thereby increasing the voltage V_m in Fig. 2.1. However the negative feedback increases the gate of M_{34} to absorb the extra PMOS current to restore the voltage V_m to V_b. Hence the PMOS (M_{33}) current reduces while the NMOS (M_{34}) current increases with an increase in the power supply voltage making the sum of the currents nearly constant. As we will see next

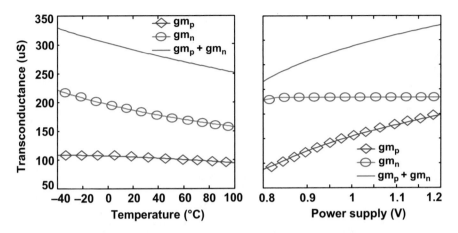

Fig. 2.3 Variation of inverter transconductance with temperature and supply

this stabilizes the NMOS + PMOS transconductance. The common-mode voltage V_{cm} is used to bias the rest of inverters in the design.

Figure 2.3 shows the variation of transconductance with temperature (left) and with supply voltage (right). We note that though the overall transconductance is not completely constant it only varies from 320 to 260 µS with a 140° change in temperature. Regarding supply variation, the NMOS transconductance is constant with power supply and is higher than the PMOS transconductance limiting the overall transconductance variation from 220 to 320 µS (37 %) with a 40 % change in power supply. The variation in transconductance with normal replica biasing [6] would have been from 180 to 361 µS (67 %) for the same conditions. As we will see later with constant-gm biasing, in the next sub-section, even these variations will be eliminated. However, before we discuss constant-gm biasing let us discuss the tradeoffs involved in the ratioing of the NMOS and PMOS transistors.

2.1.1   Optimal NMOS-PMOS Ratioing

To evaluate the tradeoffs involved in ratioing the NMOS and PMOS transistors we study the variation of the overall transconductance for different NMOS-PMOS ratios, i.e., Wp = 2 Wn, Wp = Wn, and Wp = 0.5Wn versus the power supply voltage. The variation of transconductance using SCCB is observed to be always lower than the corresponding traditional replica biasing technique. Second, as we keep reducing the PMOS size the variation of the overall transconductance with power supply is reduced. However, this reduces the overall transconductance and also demands a higher current in the PMOS of the auxiliary inverter. Hence for this and other designs the PMOS and NMOS widths are selected to be of equal size as a design compromise.

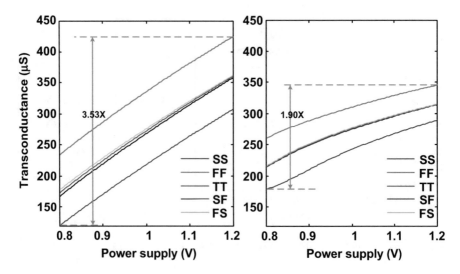

Fig. 2.4 Variation of inverter transconductances with power supply across process corner for traditional replica biased inverters and SCCB inverters

Figure 2.4 shows the transconductance variation of SCCB (right) and traditional replica biased (left) inverters with power supply across process corners. The transconductance variation of SCCB inverter is 1.9X while that for the replica biased inverter is 3.53X. This is roughly a 50 % reduction in the transconductance variation using this technique alone. All the auxiliary inverters used in the fabricated designs are biased with the voltage V_{cm}. This will make sure that all the NMOS transistors in these inverters have the same bias current of Iref. With PVT variations the NMOS transconductance remains constant but the PMOS transconductance will vary. Since we have designed the PMOS transconductance to be lower than the NMOS transconductance, the overall transconductance variation is reduced. Any mismatch between the transistors in the biasing network and the forward path will only result in an input referred offset due to feedback around the loop.

Semi constant biasing allows us to have different currents in the PMOS and NMOS transistor. This enables us to select an optimal NMOS bias current where the PMOS and NMOS nonlinearity is mutually cancelled as discussed in non-linearity cancellation section (Fig. 2.1).

2.1.2   Non Linearity Cancellation in Inverters

The sizing of these inverters are done to get the maximum open loop linearity. As it is shown later in ADC Driver chapter, the linearity of the output partially depends on the open loop linearity of the inverters. The threshold voltage (V_T) of the transistors is close to mid power supply in lower technologies. When the input voltage is high the PMOS transistor goes to subthreshold region and NMOS transistor stays in saturation. Similarly with low input voltage NMOS goes to sub threshold and PMOS stays in saturation.

2.1.3 Case 1: Small Input

When the input is small both the transistors are in saturation. The PMOS and NMOS current in saturation region with short channel effects is given by

$$I_P = \beta_p \left((V_{DD} - V_{IN} - V_{TP})^2 - \theta_p (V_{DD} - V_{IN} - V_{TP})^3 \right) \qquad (2.1)$$

$$I_N = \beta_n \left((V_{IN} - V_{TN})^2 - \theta_n (V_{IN} - V_{TN})^3 \right) \qquad (2.2)$$

$$\text{where} \quad \beta_x = \mu_x C_{ox} \frac{W_x}{L_x} \qquad (2.3)$$

where μ_n, C_{ox} are mobility of electrons, oxide capacitance and V_{TN}, V_{TP}, θ_n and θ_p are the threshold voltages and short channel parameters of NMOS and PMOS transistors respectively. The output current (I_{out}) given as the difference between the PMOS and NMOS current from Eq. (2.3) as

$$I_{out} = a_0 + a_1 V_{IN} + a_2 V_{IN}^2 + a_3 V_{IN}^3 \qquad (2.4)$$

where

$$a_0 = \left((V_{DD} - V_{TP})^2 - \theta_p (V_{DD} - V_{TP})^3 \right) \beta_p$$
$$- \left(V_{TN}^2 - \theta_n V_{TN}^3 \right) \beta_n \qquad (2.5)$$

$$a_1 = \left(2V_{TN} + 3\theta_n V_{TN}^2 \right) \beta_n$$
$$- \left(2(V_{DD} - V_{TP}) - 3(V_{DD} - V_{TP})^2 \theta_p \right) \beta_p \qquad (2.6)$$

$$a_2 = \left(1 - 3(V_{DD} - V_{TP})\theta_p \right) \beta_p - (1 - 3V_{TN}\theta_n) \beta_n \qquad (2.7)$$

$$a_3 = \theta_p \beta_p + \theta_n \beta_n \qquad (2.8)$$

For a differential implementation all the even harmonics cancel away. Since the input signal is small for this case the third order term is negligible compared to the fundamental.

2.1.4   Case 2: Large Input

When the input is large, one transistor will be saturation and other transistor will be in subthreshold region. For analysis assume the NMOS is in saturation and PMOS is in subthreshold. If the input to the inverter is V_{IN}, the short channel NMOS current in saturation region and the PMOS subthreshold current in given by

$$I_N = \beta_n \left((V_{IN} - V_{TN})^2 - \theta_n (V_{IN} - V_{TN})^3 \right) \tag{2.9}$$

$$I_P = I_o exp \left(\frac{(V_{DD} - V_{IN} - V_{TP})}{\eta U_T} \right) \tag{2.10}$$

where I_o is the minimum current in saturation region of PMOS transistor and U_T is the thermal voltage. Output current (I_{out}) is given by $I_P - I_N$.

$$I_{out} = I_o exp(\frac{(V_{DD} - V_{IN} - V_{TP})}{\eta U_T})$$
$$- \mu_n C_{ox} \frac{W}{L} \left((V_{IN} - V_{TN})^2 - \theta_n (V_{IN} - V_{TN})^3 \right) \tag{2.11}$$

Expanding in Taylor series expansion we get the output current as

$$I_{out} = b_0 + b_1 V_{IN} + b_2 V_{IN}^2 + b_3 V_{IN}^3 \tag{2.12}$$

where

$$b_0 = \left(6 + 6(V_{DD} - V_{TP}) + (V_{DD} - V_{TP})^3 \right) I_0$$
$$- \beta_n \left(V_{TN}^2 - \theta_n V_{TN}^3 \right) \tag{2.13}$$

$$b_1 = \beta_n \left(2V_{TN} + 3\theta_n V_{TN}^2 \right) - \left(6 + 9(V_{DD} - V_{TP})^2 \right) I_0 \tag{2.14}$$

$$b_2 = (3 + 3(V_{DD} - V_{TP})) I_0 - \beta_n (1 - 3V_{TN}\theta_n) \tag{2.15}$$

$$b_3 = (\theta_n \beta_n - I_0) \tag{2.16}$$

The even order harmonics cancel away in differential implementation. The coefficient of the third order term can be minimized by choosing the peak PMOS current close to the product of β_n and θ_n of NMOS thereby improving linearity.

2.1.5   Simulation

The biased inverter along with the auxiliary inverter is simulated with 450 mV peak to peak input for various bias currents. The PMOS current gets adjusted automatically set by the negative feedback. The output current is taken in a low impedance node and the intermodulation distortion is measured. As seen from Fig. 2.7, the IMD decreases with bias current, it reaches a minimum and then increases. Since the IMD is sensitive with process, the simulation is done across slow-slow (SS), slow-fast (SF), fast-slow (FS), fast-fast (FF) and typical (TT) cornors. The optimal bias current of 7 μA ensures that even across cornors inter modulation distortion is always less than -72 dB. This gives an 20 dB improvement in the open loop linearity of inverter across cornors with respect to constant voltage biasing.

2.2   Constant Current Biasing

SCCB fixes the current in one transistor (NMOS in this design) while the current in other transistor (PMOS) varies with PVT. However, the PMOS transistor current can be fixed by adjusting its source voltage as shown in Fig. 2.5. Like in SCCB, the NMOS transistor (M_{37}) in Fig. 2.5 is biased using a constant current reference (I_{ref}). The input and output voltage of the main inverter (M_{36} and M_{37}) is made equal by using an auxiliary NMOS transistor (M_{38}) and OTA_b in negative feedback. Transistor (M_{38}) creates a voltage drop across the bias resistor R_b by pulling current from the source of M_{36} to make the PMOS (M_{36}) and NMOS (M_{37}) equal to I_{ref}. The resistor

Fig. 2.5 Circuit schematic
for constant current biasing
for inverters

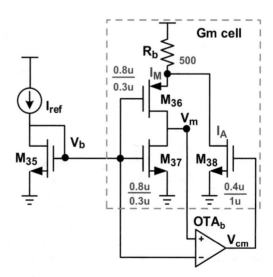

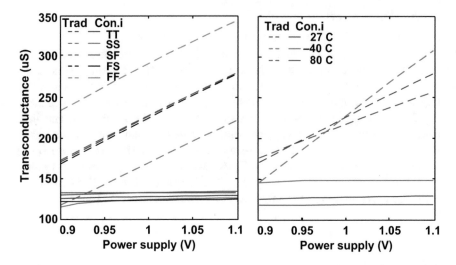

Fig. 2.6 Variation of constant current biased inverter *gm* with power supply across process corners at 27 °C and with temperature in typical corner

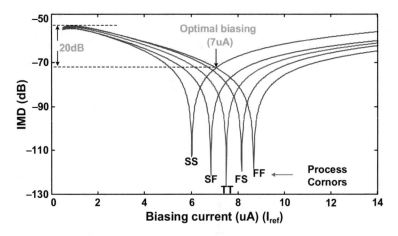

Fig. 2.7 Choice of bias current based on intermodulation distortion

R_b and transistor M_{38} are selected such that they maintain a finite non-zero voltage drop across R_b over PVT. At the nominal conditions the drop across the resistor is around 20 mV in this design. For example when the power supply increases, the negative feedback increases the voltage drop across R_b by increasing the current in M_{38}. This modulates the source voltage of M_{36} to make its current equal to I_{ref}.

Figure 2.6 show the simulated variation of the inverter transconductance with a 20 % variation in power supply voltage across process corners at 27 °C (left) and with temperature at typical corner (right). The dotted line corresponds to the variation of a traditionally biased inverter (Fig. 1.13) of the same size. The solid lines shows the transconductance variation for the constant current biasing technique. The variation of transconductance with process corners is only 9 % with constant current biasing as compared to 97 % variation with traditional inverter biasing. However, the transconductance of the constant current biased inverter and traditionally biased inverter varies by 22 and 66 % with temperature at typical process corner. This is because the mobility of the transistors changes along with the threshold voltage with temperature. Constant current biasing provides constant transconductance only with a change in the threshold voltage and not with mobility change. Hence we adopt a constant gm biasing scheme to solve the transconductance dependence on mobility as described in the next sub-section.

2.3 Constant-gm Biasing

Constant gm biasing for differential pair OTAs with PVT tolerance was proposed in [20]. We adopt the technique for inverters and update the bias current I_{ref} in Fig. 2.5 to make the transconductance PVT tolerant. Figure 2.8 shows the circuit schematic for constant gm biasing for inverters. Here I_M and I_A are the main inverter and auxiliary transistor. The negative feedback loop with OTA_1 ensures the input and output of the main inverter is equal to V by generating the bias voltage V_{cm}. If we reuse the main (I_M) and auxiliary inverters (I_A) as a gm cell in the design with I_A biased at V_{CM}, then I_M will be biased at the constant gm as in the bias network. The transistor M_{39} creates a voltage V at the gate of $M_{41,42}$ by pumping current

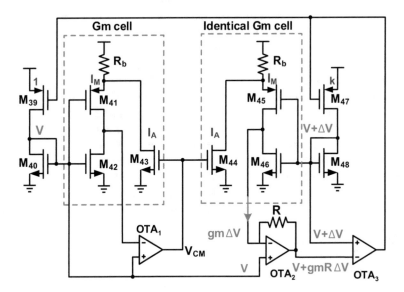

Fig. 2.8 Circuit schematic for constant gm biasing for inverters

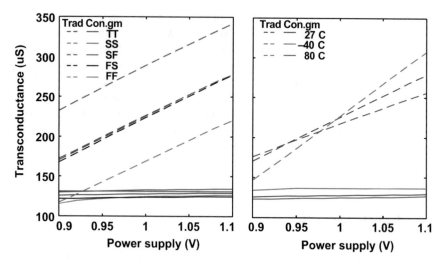

Fig. 2.9 Variation of constant gm biased inverter transconductance with power supply across process corners

into the diode connected transistor (M_{40}). Further the transistor M_{39} and M_{47} has a slight difference in their aspect ratio (1:k) to create a small voltage ΔV greater than any offset across identical gm cells ($M_{41,42}$ and $M_{45,46}$). An input of $V+\Delta V$ is given to an identical gm cell with its auxiliary biased at V_{cm}. The output current $g_m\Delta V$ is converted to a voltage $g_m\Delta VR$ using the transimpedance amplifier (OTA_2). The gate of transistor M_{39} and M_{47} are controlled using OTA_3 to make $gmR = 1$ using negative feedback. With PVT variations, the voltage V is adjusted by the bias network so that the transconductance (g_m) of the inverter remains constant at $1/R$. If the main inverter and the auxiliary transistor in the gm cell is biased with voltage V and V_{cm}, then the gm cell has the transconductance of $1/R$. To avoid the variation of R across corners ($\pm20\,\%$), the resistor R is selected to be an offchip component.

Figure 2.9 shows the simulated variation of the inverter transconductance with a 20 % variation in the power supply voltage across process corners at 27 °C (left) and with temperature at typical corner (right). The dotted line corresponds to the variation of a traditionally biased inverter (Fig. 1.13) of the same size. The solid lines shows the transconductance variation for constant gm biasing technique. The variation in transconductance reduces from 97 to 8.7 % across process corners at 27 °C and power supply with the constant-gm biasing technique. The variation in transconductance reduces from 66 to 9.2 % across temperature at 27 °C and power supply with the constant-gm biasing technique.

The constant-gm technique is an updated version of the constant current biasing technique discussed above and to a large extent solves the PVT variability of inverter based designs making them significantly more production friendly. Since replica biasing relies on the matching of the transistors, a Monte Carlo simulation was

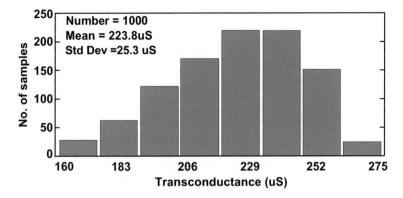

Fig. 2.10 Monte Carlo simulation for a constant gm inverter

performed on the constant gm biased inverter (size: PMOS/NMOS 1u/0.2u) as shown in Fig. 2.10. The mean and standard deviation in transconductance for 1000 runs is obtained as 223.8 and 25.3 μS.

2.4   Conclusion

Inverters has proven to have better transconductor efficiency and are inherently linear. This chapter provides a tutorial of PVT tolerant inverter based circuits. Semi constant current biasing can be employed to increase the PVT tolerance of inverter and also to increase its linearity. Constant current and constant gm biasing of the inverter further increases the PVT tolerance of the inverter. Measurements from semi constant current biased inverters in ADC Driver and tunable filter were used to verify improved PVT tolerance.

Chapter 3
Inverter Based OTA Design

The increased demand for battery operated devices has placed added pressure on lowered supply voltages. Technology scaling proportionally scales supply voltages to maintain device reliability but threshold voltages have not scaled as rapidly to limit the off current leakage in transistors. Delta sigma ADCs, particularly, continuous time delta sigma modulator are attractive due to implicit anti-aliasing, relaxed speed requirements on the active elements and the use of resistive input impedances. Although the use of multibit quantizers relaxes the design of the loop filter and are less sensitive to clock jitter, single-bit designs are simpler and do not require any dynamic element matching. The design of first integrator in a single bit modulator with adequate linearity and low power is always a challenge particularly in lower technologies [21].

In general, ADCs used in communication applications require high SFDR due to the presence of large blockers [22]. The OTA described in this paper is designed for a continuous time third order single bit delta sigma modulator for digitizing a 10 MHz signal band with an resolution of 12 bit and a minimum SFDR of 90 dB. The slew rate requirements on the OTA to be used in the third order modulator with optimized NTF zero is estimated is to be over 288 MV/s. The linearity and noise of the first integrator dictates the performance of the modulator. Hence the OTA has to have a linearity of 90+ dB, an effective SNR of 12 bit and minimum slew rate of 288 MV/s over a 10 MHz signal bandwidth. Designing traditional OTA to meet these specifications is extremely difficult with a supply voltage of 0.9 V, particularly, when the swing and linearity requirements are high. In scaled technologies, it is increasingly difficult to realize high output impedance tail current sources resulting in limited CMRR. Low voltage OTA designs have been proposed by biasing the transistors in weak inversion where linearity becomes a concern [3]. Nonlinearity cancellation techniques have been proposed in [7–10] at higher supply voltages and CLS has been used to improve the linearity of sampled data system amplifiers [23]. Non-cascoded inverter based OTA designs with common

© Springer International Publishing AG 2017
R.K. Palani, R. Harjani, *Inverter-Based Circuit Design Techniques for Low Supply Voltages*, Analog Circuits and Signal Processing,
DOI 10.1007/978-3-319-46628-6_3

mode cancellation have been proposed but have limited gain [5]. The differential pair front end OTA designs are attractive as they reject the common mode signals and amplify only the differential signals. The tail current source in differential pair gives common mode rejection by providing negative feedback only to common mode signals. This reduces the effective trans conductance of the differential pair to common mode signals, thereby attenuating the common mode components. The pseudo- differential circuits amplify both the common mode and differential signals and hence their bias voltage depends on the input common mode voltages. In this chapter we introduce common mode rejection for inverter based OTA designs by using parallel negative feedback without restricting input swing.

Cascoding is a technique used widely to increase the gain of the OTA. The idea behind the cascode structure is to convert the input voltage to a current using a common source stage and apply the result to a common-gate stage. Since common gate stage takes input at its low impedance node, the frequency response of the cascodes are better than the cascading the stages. However in lower technologies, cascoding of transistors are limited by the reduced power supplies. The cascode transistors in class A structures are biased at the same current as the common source amplifier. However biasing the transistors in class AB structures (inverters) are challenging as their bias current depends on the input common mode voltage. In this chapter, we derive the bias current of the cascode transistors in inverter from the meta stable voltage of the inverter.

Use of inverters further leads to an increased linearity and reduced input referred noise of the OTA. The inverter based OTA is fabricated in TSMC 40 nm General purpose CMOS technology with a nominal 0.9 V supply to be used in front integrator of a delta sigma modulator. For testing purpose, OTA is configured as an inverting amplifier of gain one using 3.4 kΩ resistors. The inverters in the OTA are biased using traditional replica meta stable point biasing [6]. The OTA achieves a total harmonic distortion HD of −90.6 dB over 10 MHz signal bandwidth while driving a 2 pF capacitive load. The measured input referred noise (IRN) density of the OTA is $12 \, nV/\sqrt{(Hz)}$. Further the measured low frequency common mode rejection ratio (CMRR) and power supply rejection ratio (PSRR) is obtained as 97 dB and 61 dB. The OTA is designed with sufficient margins for PVT variations. The measured third harmonic distortion varies only by 11.5 dB over a 120 °C variation in temperature and by 9 dB for 18 % variation in supply voltage.

3.1 OTA Design

The OTA consists of three stages (Fig. 3.1), the common mode rejection stage (CMRS), gain stage (GS) and the driver stage (DS). The common mode rejection stage is primarily responsible for rejecting the input common mode variations and translating the unknown input common mode voltage to a known metastable voltage of the inverters in gain stage. This stage also provides a low gain of 10 dB to reduce the input referred noise of the OTA. The gain stage is a telescopic cascoded stage and

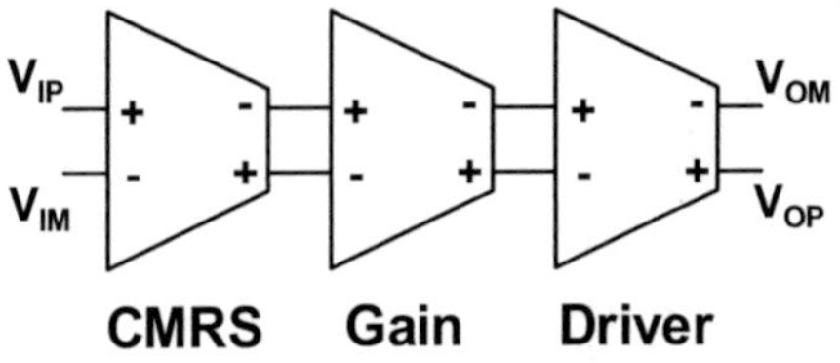

Fig. 3.1 Block diagram of
the proposed inverter based
OTA

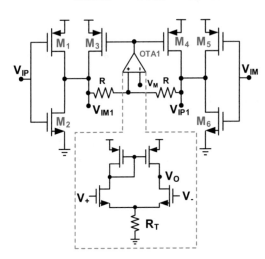

Fig. 3.2 Circuit schematic of
CMRS stage

is primarily responsible for the OTA gain. The driver stage forms the last stage of
the OTA and it is responsible for driving the resistive load of 3.4 kΩ and capacitive
load of 2 pF. The driver stage is designed with a gain of 20 dB and the gain varies
only by 2 dB across the entire OTA output swing.

3.1.1 Common Mode Rejection Stage

Figure 3.2 shows the circuit schematic of common mode rejection stage. The input
differential voltage is applied across the transistors ($M_{1,2}$ and $M_{5,6}$). The transistors
M_{1-3} are identical to M_{4-6}. The common mode voltage at the output of inverters
is sensed by the resistors R and is regulated by the transistors ($M_{3,4}$) and OTA1.
The resistor is selected such that the gain of this stage is around 10 dB. The OTA1
is realized as five transistor pair as shown in inlet of Fig. 3.2. The five transistor
pair OTA1 along with transistor $M_{3,4}$ forms a two stage OTA design with unity
gain feedback. The differential and common mode components are converted to
current by transistors ($M_{1,2}$ and $M_{5,6}$). The smaller resistor R helps in compensating
the common mode feedback loop by reducing the DC gain. The common mode
feedback results in low impedance of $1/Agm_3$ for the common mode signals, where
A is the gain of OTA1. However the feedback is broken for differential mode signals
resulting in an impedance of R. The differential and common mode gain of this stage

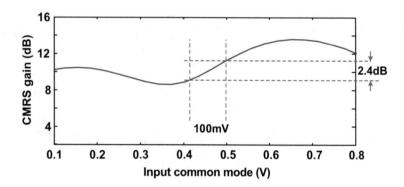

Fig. 3.3 Simulated CMRS gain with input common mode voltage

is given in Eq. (3.2). Figure 3.3 show the simulated differential gain with the input common mode voltage. The gain varies only by 2.4 dB with 100 mV variation in the common mode voltage.

$$A_{CM} \approx \frac{g_{m1} + g_{m2}}{Ag_{m3}} \tag{3.1}$$

$$A_{DM} = \frac{g_{m1} + g_{m2}}{g_{ds1} + g_{ds2} + g_{ds3}} \tag{3.2}$$

The CMRS stage provides output V_{IP1} and V_{IM1} over a known common mode voltage V_M. The voltage V_M is the meta stable voltage of the gain stage inverters (M_8, M_{13} in Fig. 3.4). Transistors (M_8, M_{13}) are sized such that V_M is close to mid supply. This allows for the cascoding of both NMOS (M_{11}) and PMOS (M_{10}) sides with sufficient overdrive voltage.

3.1.2 Gain and Driver Stage

Figure 3.4 shows the gain and driver stage of the OTA along with output common mode feedback. The output of CMRS stage over an known common mode meta stable voltage V_M is fed to the inverters (M_8, M_{13}). The inverter converts the input voltage V_{IP1} and V_{IM1} into current. The current is then converted back to voltage using the output impedance of the gain stage. The gain stage is cascoded using the transistors $M_{10,11}$ to increase the stage output impedance. This stage has a gain of 52 dB and is responsible for majority of the gain of OTA. However cascoding requires headroom requirements and hence it reduces the output swing at the gain stage. The output swing can be increased by using the driver stage. The transistors M_7 and M_{12} form the driver stage of the OTA and provides a gain of 20 dB. This reduces the output swing at the gain stage to 45 mV. The gain stage is designed such that all the transistors remain in saturation region over the entire swing of 45 mV.

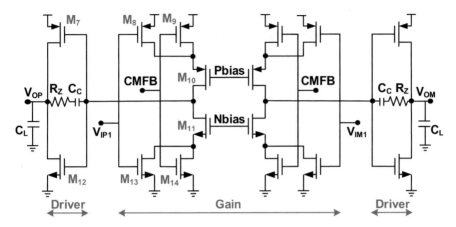

Fig. 3.4 Circuit schematic of gain and driver stage

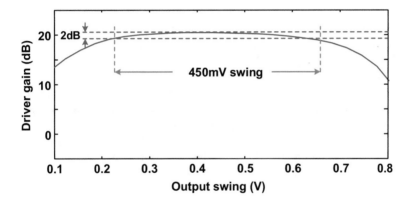

Fig. 3.5 Simulated driver gain with output swing

Figure 3.5 shows the simulated driver stage gain across maximum swing of the OTA. The gain is almost constant with only 2 dB variation across the entire output swing. The OTA is compensated using miller compensation using compensation capacitor C_c and a zero nulling resistor R_Z. The output common mode voltage is sensed by the resistors and is regulated by transistors ($M9, M14$) using a five transistor OTA stage in feedback (not shown). Because all signals are referenced to ground and have a shared path, the differential-mode and common-mode paths can use the same compensation capacitor and resistor resulting in low power operation. The operating point of the Gain and Driver stages is unaffected by the input common mode voltage due to common mode regulation by CMRS stage.

Figure 3.6 shows the biasing of transistors in the gain stage of OTA. The transistors $M_{15,16}$ in Fig. 3.6 are identical to the gain stage transistors $M_{8,13}$ and hence it generates the meta stable voltage (V_M) of the gain stage inverters. The common mode rejection stage translates the input from unknown input common

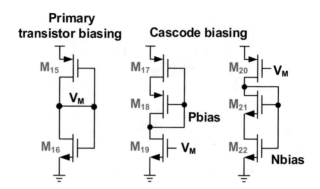

Fig. 3.6 Biasing of transistors in gain stage

mode voltage to this meta stable voltage V_M. Hence the input to gain stage is always at its meta stable voltage where the transconductance is maximum. The bias current in the gain stage is determined by its meta stable voltage rather than any input common mode voltage. This allows cascode biasing at the current determined by V_M as shown in Fig. 3.6. The cascode transistors carry the current from the gain transistors (M_8, M_{13}) and CMFB transistors (M_9, M_{14}). The cascode biasing transistor M_{18} and M_{21} are identical to the cascodes (M_{10} and M_{11}). The transistors $M_19, 20$ are the bias current sources derived from the metastable voltage V_M. These current sources are used in low voltage cascode biasing as shown in Fig. 3.6. Further the aspect ratio of the driver stage inverters are scaled version of the gain stage inverters. The output common mode feedback loop maintains the output common mode voltage to the metastable voltage V_M by adjusting the gates of M_9 and M_{14}. The OTA is Miller compensated with the unity gain frequency approximately at $(gm_8 + gm_{13})/C_C$ and the location of the second pole at $(gm_7 + gm_{12})/C_L$. Both power supply and temperature variations affect the transconductances ($(gm_8 + gm_{13})$ and $(gm_7 + gm_{12})$) in the same direction. The phase margin of the OTA is always greater than 60° over PVT as both the second pole and the UGB are affected in same direction. The OTA is designed to be well compensated from −40 to 120 °C and can handle power supplies from 0.84 to 1 V.

3.2 Measurement Results

The OTA was fabricated in TSMC's 40 nm general purpose technology and occupies an area of 0.0025 mm^2 as shown in Fig. 3.7. The OTA was configured as an inverting amplifier of gain one with resistors 3.4 kΩ and loaded with the output capacitance of 2 pF. Phase and amplitude matched 50 Ω baluns were used to interface with the single ended equipment. Sharp passive band pass filters were used to clean the harmonics generated by the signal generators. Figure 3.8 shows the measurement setup for OTA testing. The OTA drives the spectrum analyser

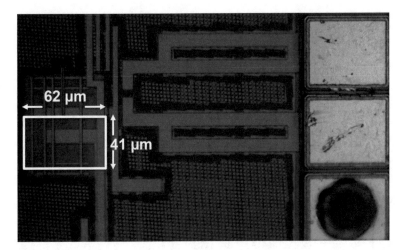

Fig. 3.7 Micrograph of proposed OTA

Fig. 3.8 Test setup of the OTA

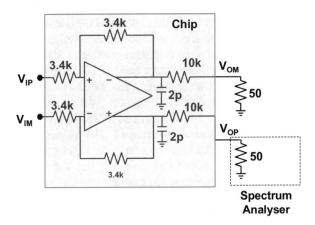

through a $10\,k\Omega$ resistor to avoid its $50\,\Omega$ loading. This results in a signal attenuation which is calibrated from the measurement results. Figure 3.9 shows the magnitude response of the inverting amplifier. The flat magnitude response without peaking indicates a well compensated system. The 3 dB frequency of the inverting amplifier corresponds approximately to the unity gain frequency of the loop gain and is obtained as 502 MHz. The loop unity gain frequency $\omega_{u,loop}$ can be expressed in terms of OTA unity gain frequency ω_u for inverting amplifier as

$$\omega_{u,loop} = \omega_u \frac{1}{1 + A_{CL}} \tag{3.3}$$

where A_{CL} is the closed loop gain of the amplifier ($A_{CL} = 1$ in this case). Hence the unity gain frequency of the OTA is estimated to be 1 GHz. The slew rate of

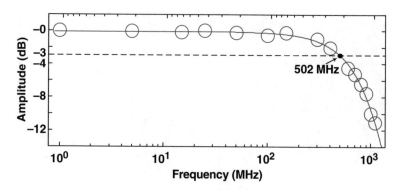

Fig. 3.9 Measured magnitude response of the OTA

inverter based amplifier (class AB) are typically higher than corresponding class A amplifiers due to its push pull operation. Figure 3.10 shows the screen shot of the slew rate measurement of OTA. A fast step with step size 500 mV is given at the OTA input as shown in yellow color and the rise time of the output is measured. The rise time and fall time of the output is measured as 600 and 940 ps which translates to the average positive and negative slew rate of 360 and 265 V/μs. The step response shows no ringing confirming a good phase margin in the system.

Figure 3.11 shows the measured CMRR and PSRR of the OTA. Low frequency CMRR is measured as 97 dB and the 3 dB point is at 2.5 MHz. PSRR is 61 dB till 30 MHz and the 3 dB point is at 35 MHz. Figure 3.12 shows the screenshot of the single ended measured spectrum of OTA output at 9.5 MHz. Second harmonic is present partially due to its content in the input spectrum and also its a single ended measurement. The third harmonic falls lies within the 3 dB bandwidth (502 MHz) of the inverting amplifier. The third harmonic distortion at 9.5 MHz, 900 mV$_{pp,diff}$ is −91.9 dB while operating from 0.9 V supply.

OTA is designed such that the transistors are in saturation over PVT. Further all inverters are biased at their meta stable point and hence all transconductance varies in same direction. PVT insensitivity is verified by measuring third harmonic distortion with frequency over power supply and temperature ranges. Figures 3.13 and 3.14 shows the measured third harmonic distortion with frequency over temperature and power supply. At 9.5 MHz, the variation in third harmonic distortion is 11.5 dB with 120 C variation in temperature. Further IMD variation is 9 dB with 18 % variation in power supply. When the power drops below 0.84 V, the cascode current source transistors (M_{10}, M_{11} in Fig. 3.4) enter into linear region, dropping the loop gain and hence the linearity. Further inverters has higher open loop linearity than the conventional differential pair at the same transconductance. This is because of current limiting in the differential pair and also the body effect incurred by the input transistors. Signal generator limits the input signal to have the linearity of 92 dB hence at low frequency we get the maximum linearity to 90 dB. Further since the 3 dB bandwidth of the OTA is 50x larger than the operating frequency, the third

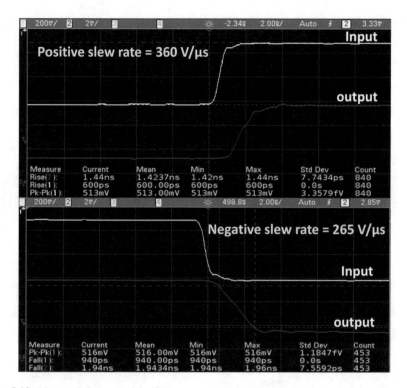

Fig. 3.10 Measured slew rate of the OTA

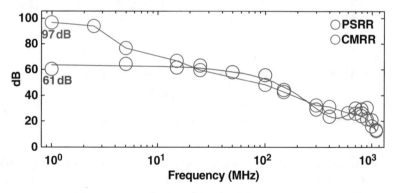

Fig. 3.11 Measured common mode rejection ratio (CMRR) and power supply rejection ratio (PSRR) of OTA

harmonic component will fall well within the band. The 3 dB frequency of the OTA is 50x larger primarily to have the higher loop gain at the operating frequency to get the desired linearity. All the transistors in inverter based amplifiers contribute to both signal and noise and hence it has superior noise performance than differential pair based amplifiers. Further gain in CMRS stage suppresses the noise contribution

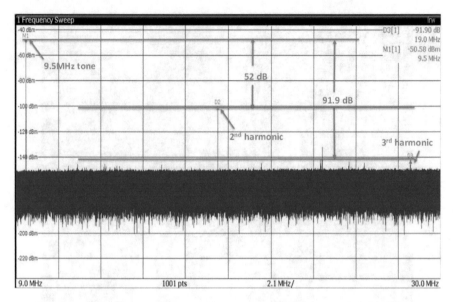

Fig. 3.12 Screen shot of single ended measured spectrum of OTA output at 9.5 MHz 900 mV_{ppdiff}

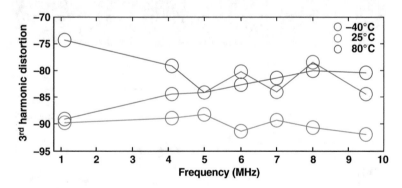

Fig. 3.13 Measured third order distortion versus frequency over temperature

from gain and driver stage. The measured integrated noise over 10 MHz bandwidth is 82 μV rms including the feedback network and is equal to 37 μV rms for the OTA only, which translates to an SNR of 78.7 dB for a $V_{pp-diff} = 900$ mV.

Table 3.1 shows the performance summary and comparison with prior art. Although this design is an inverter based design, we obtain low frequency PSRR of 61 dB and CMRR of 97 dB. Although [5] has the lowest power supply, the design is biased in weak inversion limiting signal bandwidth and linearity. Our work has the lowest power supply with all the transistors in strong inversion. We obtain the best total harmonic distortion of −90.6 dB with an output swing of 0.9 Vpp while driving 2 pF and 3.4 kΩ load. Further the third harmonic distortion varies only by 11.5 dB over 120 °C variation in temperature and by 9 dB for 18 % variation in power supply.

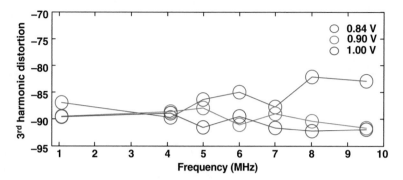

Fig. 3.14 Measured third order distortion versus frequency over power supply

Table 3.1 Summary and comparison with prior art

	[7]	[8]	[5]	[3]	This work
Technology (nm)	180	350	180	180	40
Supply (V)	0.5	3.3	1.5	1.8	0.9
Power (mW)	0.08	10.5	9.5	5	1.1
Signal BW (MHz)	0.14	20	40	100	10
Noise (nV/$\sqrt{\text{Hz}}$)	70	149	23	–	12
Output swing ($V_{pp\text{-}diff}$)	0.7	1.3	0.9	1.6	0.9[a]
THD (dB)	−57[a]	−69	−60	–	−90.6

[a]THD is estimated from IM3

3.3   Conclusion

This chapter describes a completely inverter based OTA design. Although this design is an inverter based design, we obtain low frequency PSRR of 61 dB and CMRR of 97 dB. This work has the lowest power supply with all the transistors in strong inversion. We also obtain the best total harmonic distortion of −90.6 dB with an output swing of $V_{DD}/2$ (450 mV) single ended while driving 2 pF capacitive load and 3.4 kΩ resistive load. Further the third harmonic distortion varies only by 11.5 dB over 120 10 °C variation in temperature and by 9 dB for 18 % variation in power supply.

Chapter 4
ADC Driver

The demand for hand-held mobile phones and their longer battery life-time motivates low-voltage integrated circuit design. Mobile video and high-definition television applications require high-speed, medium resolution, and low-power design specifications, which are the realm of pipelined and SAR ADCs. The SAR ADC has become increasingly popular as it is based on switching rather than amplifying, contains less hardware than other topologies and it scales well with technology. However, due to its inherent high input capacitance the design of the input driver is becoming more difficult, especially at lower technology nodes [24]. Further, the design of rail-to rail output amplifiers is also becoming increasingly difficult at lower technologies due to poor output impedance and higher threshold voltages. Logic circuits' static power dissipation limit the scaling of threshold voltages of transistors. The noise requirements of an operational transconductance amplifier (OTA) design becomes increasingly more stringent for lower input voltage swings. Linearity requirements demand high overdrive voltages. On the other hand high overdrive voltages result in higher noise for the same current.

Rail-to-rail inputs for an ADC reduce its power consumption while simultaneously increasing the dynamic range. Ring amplifiers [25] are a topology that can give rail-to-rail swing at the cost of increased noise. Passive amplification techniques [24, 26] have demonstrated an ability to amplify signals to rail-to-rail levels at the cost of increased loading on the driver. An ideal ADC driver should be able to amplify the input signals to rail-to-rail voltages and sample the signal onto the input capacitor of the ADC. The requirements on the OTA gain-bandwidth product (GBP) and DC gain keep increasing with increased closed loop gain. Inverter-based designs are becoming the preferred topologies at lower technologies nodes due to their higher g_m/I_D efficiency [27].

In this work we present an ADC driver which employs a current-mirror based architecture. Voltage-to-current conversion is done using a passive device and a negative feedback circuit. The current is amplified by a current-mirror [17].

© Springer International Publishing AG 2017
R.K. Palani, R. Harjani, *Inverter-Based Circuit Design Techniques for Low Supply Voltages*, Analog Circuits and Signal Processing,
DOI 10.1007/978-3-319-46628-6_4

A trans-impedance amplifier (TIA) converts this current back to a voltage with half V_{DD} swing. A passive RC circuit is used as a first-order anti-alias filter (AAF) before sampling is performed. Passive amplification is employed to amplify the signals to rail-to-rail levels [26]. We also show that the gain bandwidth product (GBP) and the DC gain of the negative feedback loop are not affected by the closed-loop gain. Inverters are used as amplifiers in this design and the NMOS devices in the inverters are biased at a semi constant current biasing resulting in low PVT variations.

4.1   ADC Driver

An ideal ADC driver should be able to amplify the input signal to dynamic range of the ADC and sample onto its input capacitance. The amplifier used to drive an ADC can be in discrete time or continuous time. Further it can be open loop or a closed loop system. An open loop amplifier typically consumes lower power and can operate at higher frequency. However it is typically more nonlinear than closed loop system as the input is taken across the active elements. On the other hand closed loop system suppresses the inherent active elements non linearity by its loop gain. Further the loop gain and unity gain frequency of closed loop system increases with the closed loop gain and the bandwidth.

Lets assume an ideal amplifier driving the load capacitor C_L. The amplifier is designed to be two stage OTA with unity gain feedback. At lower power supplies, typically an OTA is used for high gain rather than opamp due to swing requirements. Hence we can either drive the load directly or drive through resistor and each has its merits as described below

4.2   OTA Driving Load

4.2.1   Driving Load Capacitor Directly

Figure 4.1 shows single ended schematic of the ADC Driver. Here C_L is the input capacitance of the ADC. The OTA in Fig. 4.1 is assumed to be a two stage design for analysis. In order to evaluate the loop gain and stability, loop in Fig. 4.1 is broken at the OTA input as shown in Fig. 4.2. Here g_{ds1} and g_{ds2} are the finite output impedances of the two stages, C_{p1} is the parasitic capacitance at the output of first stage, C_{p2} is the sum of the input capacitance of first stage (g_{m1}) and parasitic capacitance of the second stage (g_{m2}) and C_L is the load capacitance. The loop gain can be derived as

$$\frac{V_o}{V_T} = \frac{g_{m1}g_{m2}}{g_{ds1}g_{ds2}} \frac{1}{(1 + \frac{s}{P_1})(1 + \frac{s}{P_2})} \qquad (4.1)$$

Fig. 4.1 ADC driver

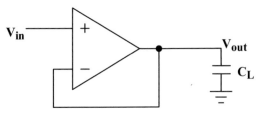

Fig. 4.2 Loop gain of the
ADC driver

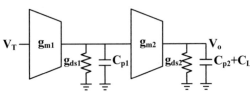

where

$$P_1 = -\frac{g_{ds1}}{C_{p1}} \qquad P_2 = -\frac{g_{ds2}}{C_{p2} + C_L} \tag{4.2}$$

Typically the poles are close to each other thereby reducing the phase margin and stability. Hence compensation techniques like miller compensation, feedforward compensation, feedback zero compensation etc are used to improve the stability. The power efficient compensation techniques like feedforward compensation gives rise to pole zero doublets in sampled data systems.

4.2.2 Driving Load Capacitor Through Resistor

Lets consider the case of driving the input capacitance of an ADC through a resistor as shown in Fig. 4.3. The bandwidth $R_f C_L$ is assumed to be higher than the signal bandwidth so that signal is not attenuated by this passive filter. The OTA here is assumed to be identical to the OTA used in Fig. 4.1. In order to evaluate the loop gain and stability, loop in Fig. 4.3 is broken at the OTA input as shown in Fig. 4.4. The loop gain can be derived as

$$\frac{V_o}{V_T} = \frac{g_{m1}g_{m2}}{g_{ds1}g_{ds2}} \left(\frac{1}{1 + s\frac{C_{p1}}{g_{ds1}}} \right)$$

$$\times \left(\frac{1 + sR_f C_L}{s^2 \frac{R_f C_L C_{p2}}{g_{ds2}} + s\left(R_f C_L + \frac{C_{p2} + C_L}{g_{ds2}}\right) + 1} \right) \tag{4.3}$$

Fig. 4.3 ADC driver

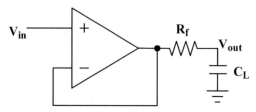

Fig. 4.4 Loop gain of the
ADC driver while driving
capacitive load through
resistor

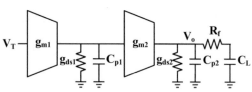

The system has three poles and one zero which can be approximated by

$$Z_{1R} = -\frac{1}{R_f C_L} \qquad P_{1R} = -\frac{g_{ds1}}{C_{p1}} \tag{4.4}$$

$$P_{2R} \approx -\frac{g_{ds2}}{C_{p2} + C_L(1 + R_f g_{ds2})} \tag{4.5}$$

$$P_{3R} \approx -\frac{C_{p2} + C_L(1 + R_f g_{ds2})}{R_f C_L C_{p2}} \tag{4.6}$$

The load capacitance and the series resistance creates left half plane zero Z_{1R}. Pole P_{1R} is created at the output of the first stage due to finite output impedance of the g_{m1} stage and associated parasitic capacitance. The second stage along with R_f and C_L creates two poles which can be approximated at P_{2R} and P_{3R}. If the load capacitance is assumed to be much greater than the parasitics (C_{p2}) and $R_f g_{ds2}$ is less than one, poles P_{2R} and P_{3R} can be further approximated as Eq. (4.7).

$$P_{2R} \approx -\frac{g_{ds2}}{C_L} \quad P_{3R} \approx -\frac{1}{R_f C_{p2}} \tag{4.7}$$

Poles P_{2R} and P_{3R} are low and high frequency poles respectively. Pole P_{3R} can be moved out of the unity gain bandwidth (ω_u) by appropriately choosing R_f thereby leaving two poles (P_{1R} and P_{2R}) and one zero (Z_{1R}) within ω_u as shown in Fig. 4.5. The zero created by R_f compensates the system and unity gain frequency of the system can be derived approximately as

$$\omega_u \approx \frac{g_{m1} g_{m2} R_f}{C_{p1}} \tag{4.8}$$

The unity gain frequency (Eq. (4.8)) to the first order depends only on the parasitic capacitance at the output of the first stage and not on the load capacitance.

Fig. 4.5 Bode plot of loop
gain of ADC driver

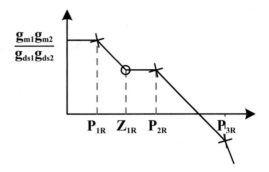

This is because the load capacitance creates both pole and zero close to each other. The series resistor R_f is selected such that there is no significant signal attenuation on the load capacitor and hence it limits the unity gain frequency (Eq. (4.8)) of the system. Further as the system (Eq. (4.3)) is self compensated no external compensation is needed and hence the driver power is lowered. The series resistor R_f further filters the noise away from the signal bandwidth as explained in Sect. 4.3.

4.2.2.1 Effect of Pole Zero Doublets

The system (Eq. (4.3)) is self compensated with left half plane zero at the load. This created a pole zero doublet in the closed loop transfer function. However since the output voltage is across the load capacitor rather than across the series resistor R_f and capacitor C_L, it creates a pole exactly at the same location as the zero Z_{1R}. Hence there is no pole-zero doublet in the closed loop transfer function from V_{in} to V_{out} thereby not affecting the settling time of the amplifier. The zero Z_{1R} helps only in the compensation of the loop and does not affect the final settling of the output on the ADC input capacitance. The closed loop transfer function can be derived for ADC driver shown in Fig. 4.3 as

$$\frac{V_o}{V_{in}} = \frac{A_{dc}}{aks^3 + s^2(bk + a) + s(k + b + \frac{A_{dc}}{Z_{1R}}) + A_{dc} + 1} \tag{4.9}$$

where

$$A_{dc} = \frac{g_{m1}g_{m2}}{g_{ds1}g_{ds2}} \qquad k = \frac{C_{p1}}{g_{ds1}} \tag{4.10}$$

$$a = \frac{R_f C_L C_{p2}}{g_{ds2}} \qquad b = R_f C_L + \frac{C_{p2} + C_L}{g_{ds2}} \tag{4.11}$$

The closed loop poles can be approximated as

$$P_{C1} \approx \frac{g_{m1}g_{m2}}{C_L\left(g_{ds1} + g_{m1}g_{m2}R_f\right)} \approx \frac{1}{R_f C_L} \tag{4.12}$$

$$P_{C2} \approx \frac{g_{ds1}C_L\left(A_{dc}R_f g_{ds2} + 1\right)}{R_f C_L\left(C_{p1}g_{ds2} + C_{p2}g_{ds1}\right) + C_L C_{p1}}$$

$$\approx \frac{g_{m1}g_{m2}R_f}{C_{p1}} \tag{4.13}$$

$$P_{C3} \approx \frac{1}{R_f C_{p2}} \tag{4.14}$$

The low frequency pole P_{C1} is at the cut off frequency of the series resistor R_f and capacitor C_L and the next pole is at the unity gain frequency (ω_u) of the loop gain (Eq. (4.3)). The high frequency pole P_{C3} is dominated approximately by the series resistance and the parasitic capacitance of the second stage. The 3 dB frequency of the closed loop response is thus dominated by the low frequency pole (P_{C1}) i.e. by the series combination of load capacitance (C_L) and the resistor (R_f).

4.3   Continuous and Discrete Time ADC Driver

The amplifier for driving the ADC can be either in the continuous time or in discrete time. The series resistance R_f is used in both the designs for comparison. Input is sampled over the ADC input capacitance C_L using the sampling clock ϕ. The series resistance R_f helps in compensation as explained in Sect. 4.2 and also filters the out of band noise which will be explained in this section.

4.3.1   Continuous Time Driver

Figure 4.6 shows the circuit schematic of continuous time driver. The closed loop gain of the driver is assumed to be A. The load capacitance C_L is driven through series resistance R_f. The resistors and the OTA contribute to the noise sampled on the capacitance.

4.3.1.1   Noise Analysis

Input resistors (R), feedback resistor (AR), series resistor (R_f) along with the switch and the operational amplifier contribute noise at the output. The noise power spectral density of the components is given by

Fig. 4.6 Continuous time
ADC driver

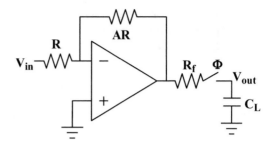

$$v_{nR}^2 = 4kTR \qquad v_{nAR}^2 = 4kTAR \tag{4.15}$$

$$v_{nRf}^2 = 4kTR_f \qquad v_{nOTA}^2 = \frac{4kT\gamma}{g_m}(1 + \eta) \tag{4.16}$$

The noise power spectral density at the output is given by

$$v_{nout}^2 = 8kTAR + \frac{4kT\gamma}{g_m}(1 + \eta)(1 + A) + 4kTR_f \tag{4.17}$$

The closed loop bandwidth of the system has a 3 dB bandwidth of $1/R_f C_L$
(Eq. (4.14)) and the integrated noise over the bandwidth is given by

$$v_{nint,out,CT}^2 = v_{nout}^2 \int_{-\infty}^{\infty} \frac{1}{1 + R_f^2 C_L^2 \omega^2} df \tag{4.18}$$

$$v_{nint,out,CT}^2 = \frac{kT}{C_L}\left(1 + \frac{2AR}{R_f} + \frac{\gamma}{g_m R_f}(1 + \eta)(1 + A)\right) \tag{4.19}$$

The noise contributed by series resistor R_f is independent of the value of resistor
R_f. This is because as the value of R_f is increased, there is a reduction in the noise
bandwidth and corresponding increase in the noise power spectral density leading
to the same integrated noise. The gain bandwidth product of the negative feedback
loop is much greater than the signal frequency resulting in better non linearity
suppression up to the band-edge frequency. If this signal is sampled directly at
nyquist all the noise past the Nyquist signal bandwidth aliases into the signal band.
However the resistor R_f limits the noise bandwidth to be $1/R_f C_L$ thereby filtering
the noise of the input resistor, feedback resistor and also the OTA before sampling.

4.3.1.2   OTA Design

If the unity gain bandwidth (UGB) of the OTA is assumed to be $\omega_{u,ota}$, the UGB of
the loopgain ω_u is given by

$$\omega_u = \omega_{u,ota} \frac{1}{A + 1} \tag{4.20}$$

Finite ω_u affects the settling behaviour of the input. The driver for N bit ADC has be settling at least to N bit accuracy. Assuming first order settling the minimum unity gain frequency required for N bit settling

$$\omega_{u,min} = 0.69N\frac{f_s}{2} \qquad (4.21)$$

$$\omega_{u,ota} = 0.69(A+1)N\frac{f_s}{2} \qquad (4.22)$$

where f_s is the sampling frequency. With increasing closed loop gain (A), the unity gain frequency requirement of the OTA is increases as shown in Eq. (4.22). This is because higher closed loop gain results in higher attenuation in the feedback factor or the loop gain. However if the closed loop gain is made independent of the feedback factor, then the unity gain frequency of the ota ($\omega_{u,ota}$) can be relaxed as explained in Sect. 4.5.

4.3.2   Discrete Time Driver

Figure 4.7 shows the circuit schematic of discrete time driver. The closed loop gain of the driver is assumed to be A. Similar to continuous time equivalent (Fig. 4.6), load capacitance C_L is driven through series resistance R_f. The switches, resistor (R_f) and the OTA contribute to the noise sampled on the load capacitance.

4.3.2.1   Noise Analysis

The switches (whose resistance R_s), the series resistor R_f, the sampling switch and the operational amplifier contribute noise at output. During the sampling phase (ϕ_b), input is sampled onto the capacitor (AC) and the feedback capacitor is reset. During this phase the capacitor AC has the noise voltage of

$$v_{nint,sam} = \frac{kT}{AC} \qquad (4.23)$$

Fig. 4.7 Discrete time ADC driver

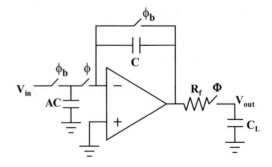

During the amplification phase the operational amplifier, switch and resistor R_f noise is sampled on the load capacitance after being filtered by the resistor R_f. The total integrated noise in the amplification phase is given by

$$v_{nint,amp} = \frac{kT}{C_L}\left(1 + \frac{\gamma(1+\eta)}{g_m R_f}(1+A) + A\frac{R_s}{R_f}\right) \qquad (4.24)$$

The total noise obtained as the sum of Eqs. (4.23) and (4.24) is given by Eq. (4.25).

$$v_{nint,out,DT} =$$

$$\frac{kT}{C_L}\left(1 + \frac{C_L}{C} + \frac{\gamma(1+\eta)}{g_m R_f}(1+A) + A\frac{R_s}{R_f}\right) \qquad (4.25)$$

Comparing the total noise with continuous time equivalent, discrete time system has more noise for higher load capacitance and higher sampling frequency. while the noise of OTA, feedback element and series resistor noise remains same for both the systems, there is difference in the noise of the input element. The noise of the input resistor (R) in continuous time is filtered by the series resistor R_f and load capacitance C_L before sampling while the noise of the switched capacitor resistor is sampled before filtering by R_f and C_L (Eq. (4.23)).

4.3.2.2 OTA Design

Similar to the continuous time equivalent the operational amplifier's minimum unity gain frequency, $\omega_{u,ota}$ required for N bit accuracy in settling is given by

$$\omega_{u,ota} = 0.69(A+1)N\frac{f_s}{2} \qquad (4.26)$$

The requirements on the unity gain frequency is increased with increase in the closed loop gain.

4.4 Simulation to Verify Noise Filtering

In order to verify the noise filtering by the series resistor R_f, simulation is performed using ideal voltage controlled voltage source to isolate the compensation effect of R_f. In this simulation, only the input resistor (1K) and the feedback resistor (1K) contribute noise. Figure 4.8a, b shows the simulation test bench of the ideal voltage controlled voltage source driving load capacitance ($C = 4\,pF$) with and without series resistor R_f. The signal bandwidth of the $R_f C$ bandwidth are 50 and 66 MHz respectively in this simulation.

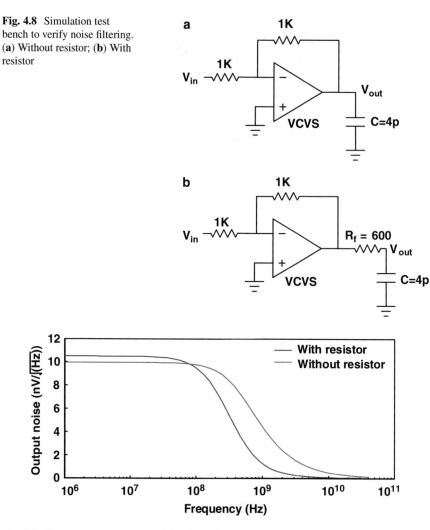

Fig. 4.8 Simulation test bench to verify noise filtering. (a) Without resistor; (b) With resistor

Fig. 4.9 Output noise power spectral density with and without series resistor R_f

Figure 4.9 shows the output noise power spectral density at the output of the driver. Blue and red colours indicate the noise with and without resistor respectively. The inband noise is higher for the driver with resistor due to additional noise component (R_f) in the system. However the out of band noise is higher for the driver without resistor as it filters the amplifier noise. Figure 4.10 shows the running integral of the output noise with frequency. The driver with resistor has 2.65X lower integrated noise compared to the driver without resistor.

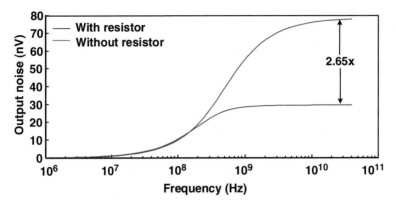

Fig. 4.10 Cumulative noise integral with and without series resistor R_f

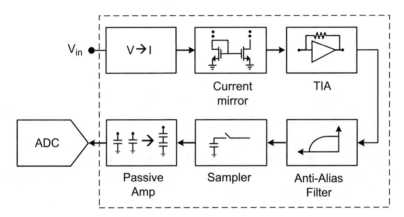

Fig. 4.11 Block diagram of the rail-to-rail output sampled ADC driver

4.5 ADC Driver Architecture

The architecture of the proposed ADC driver is shown in Fig. 4.11. The proposed ADC driver exploits the current mirror for linear amplification. The input voltage is converted to a current using a passive resistor and negative feedback. This linear current is then fed to the current-mirror where amplification is performed. The amplified current is then converted back to voltage by a trans-impedance amplifier with a half V_{DD} swing. A passive anti-alias filter is used to filter the noise above the 3-dB bandwidth of the trans-impedance amplifier after which sampling is performed. The final amplification to rail-rail swing is done passively for better linearity. For an amplifier with perfect efficiency the power required to charge a capacitor is

$$P = \tfrac{1}{2}CV^2f \qquad (4.27)$$

It can be seen from Eq. (4.27) the power required to charge a capacitor (C) to a voltage V is the same as charging a capacitor ($4C$) to $V/2$. However, charging the capacitor to $V/2$ level gives a linearity advantage to the amplifier. This principle is used and then followed by passive amplification to result in a rail-to-rail output swing which is ready for the ADC (or is the SAR ADC capacitor itself).

4.6 Components of the ADC Driver

Figure 4.12 shows the detailed circuit diagram for the ADC driver, the V-to-I, the current-mirror, the TIA, the RC AAF, the passive-amplifier and sampler that were shown in Fig. 4.11. The actual design implemented is differential in nature.

4.6.1 Current Mirror Design

The voltage to current conversion is done using negative feedback as shown in Fig. 4.12. The OTA$_1$ maintains the node V_x at the meta stable voltage V_M of biased inverter (ac ground). As a result the input voltage is converted to a current using a passive resistor R. The resistor current is absorbed by the inverter of size 1 by changing its gate voltage appropriately. The inverter of size m also has the same gate voltage and hence it absorbs a current of m times the input current. As a result we get an amplification by a factor of m in current mode. OTA$_1$ is designed as a cascade of two semi constant current biased inverters as shown in Fig. 4.14. All the inverters used in this design are multiples of the unit semi constant current biased inverters and therefore track with the PVT variations. Inverters have the advantages of the best g_m/I_D efficiency while supporting a high linear operating range for an appreciably large voltage swing.

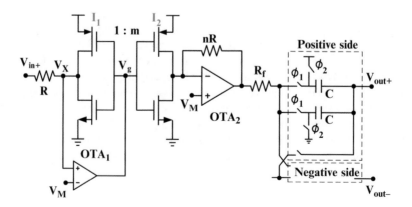

Fig. 4.12 Circuit schematic for the ADC driver

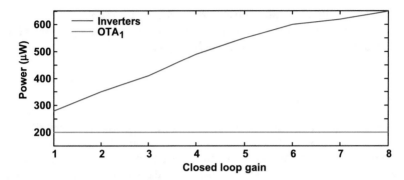

Fig. 4.13 Simulation of the voltage to current converter circuit over different closed loop gain

The OTA_1 along with the inverter I_1 forms a negative feedback loop which is independent of the closed loop gain apart from the loading caused by the gate capacitance of the inverter of size m. However, unlike inverting amplifier where the unity gain frequency and the DC gain scale with the closed loop gain, the loop parameters do not significantly depend on the closed loop gain. Hence the loop has to be designed only for gain 1 and the closed loop gain can be obtained by increasing the number of inverters (m), without significantly affecting the stability and power of the loop. This facilitates the design of a negative feedback loops at higher frequency. This is further verified by simulating the power at different closed loop gain factors, m and the results are plotted in Fig. 4.13. The blue and green colors show the power of the inverters with sizes 1 and m and OTA_1. The power of the inverter increases as the number of fingers increases with closed loop gain. The OTA_1 power remains approximately constant in this simulation.

4.6.1.1 Noise Analysis

The input resistor R, OTA_1 and the inverters contribute to the noise. The noise power spectral density of resistor R, OTA_1 and inverters are given by 4kTR, 4kT $\gamma g_{m,\mathrm{inv}}$ and 4kT $\gamma g_{m,OTA_1}$ respectively. The input referred noise of the voltage to current conversion circuit is given by Eq. (4.28).

$$v_{n,1}^2 = 4kTR \left(1 + \frac{\gamma g_{m,\mathrm{inv}} R}{1 + \frac{1}{m}} + \frac{\gamma(1 + \eta)}{g_{m,OTA1} R} \right) \tag{4.28}$$

The noise of the input resistor R and inverter noise adds directly at the input. The noise of the output inverter (size m) and OTA_1 noise get divided by the closed loop gain (m) and $g_{m,OTA1} R$ and appear at the input. As seen in Eq. (4.28), the transconductance of the inverter ($g_{m,\mathrm{inv}}$) must be small and that of the OTA ($g_{m,OTA1}$) must large for low noise. Hence, the input inverter is designed to have a lower aspect ratio thereby consuming lower current. Since gain is obtained by mirroring this inverter, the overall power is also minimized. The OTA_1 is designed for a higher trans-conductance for low noise and high loop gain.

4.6.1.2   Non Linearity Analysis

Non-linearity in the current mirror is due to g_m and g_{ds} non-linearity of the transistors. The OTA_1 in negative feedback adjusts the gate voltage of inverter I_1 to absorb the input current. The input current is created by using a passive resistor and hence it is linear. If the transconductance of the inverter is assumed to have a nonlinear function $f(x)$, then the gate voltage of the inverter (I_1) can be expressed as

$$V_g = f^{-1}\left(\frac{V_{in}}{R}\right) \tag{4.29}$$

Since the inverters I_1 and I_2 share the same gate voltage, the output current is the amplified version of the input current and is given by

$$I_o = mf(V_g) = mf\left(f^{-1}\left(\frac{V_{in}}{R}\right)\right) = m\frac{V_{in}}{R} \tag{4.30}$$

Any g_m nonlinearity in the input inverter is reflected by a change in the output voltage of OTA_1. As the output inverter has the scaled size (m) of the input inverter, any g_m nonlinearity that is introduced gets canceled. The output current is taken in a low impedance node due to the trans-impedance amplifier. Since both the drains of the input and output inverters are held at the same common mode voltage (V_M), the differences in their respective g_{ds} non-linearity is greatly reduced. Additionally, differential or pseudo-differential implementations suppress even order harmonics. The effectiveness of this non-linearity cancellation relies on the matching of the transistors and the loop gain of negative feedback loop. Therefore, these inverters are carefully layed out using common-centroid techniques to reduce any systematic mismatch. The random mismatch can be reduced by increasing the sizes of the transistors.

4.6.2   Trans-Impedance Amplifier (TIA) Design

The trans-impedance amplifier is used to convert the output current from the current mirror to voltage. It is designed using three biased inverters as OTA_2 (Fig. 4.14) with a feedback resistor. The amplified current from the current mirror is converted back to voltage by the feedback resistor (nR). Since the current-mirror inverter providing the input is designed for low bias current (for low noise purposes), it has a higher output impedance than the feedback resistor (nR) of the trans-impedance

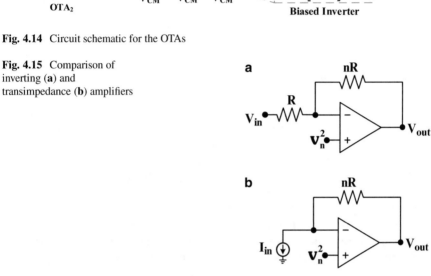

Fig. 4.14 Circuit schematic for the OTAs

Fig. 4.15 Comparison of inverting (**a**) and transimpedance (**b**) amplifiers

amplifier. The loop gain of an ideal trans-impedance amplifier is the loop gain of the operational amplifier itself as the feedback factor is one. Hence to first order, the loop gain of the transimpedance amplifier is independent of the closed loop gain. In reality the finite output impedances of the inverter I_2 reduces the feedback factor slightly.

4.6.2.1 Noise Analysis

The trans impedance amplifier differs slightly from the inverting amplifier in terms of feedback factor. Figure 4.15 shows the comparison of a trans-impedance amplifier and an inverting amplifier. The operational amplifier is assumed to be identical in both the cases for comparison purposes. The OTA noise ($v_{n,2}^2$) is amplified by the gain of $n + 1$ in an inverting amplifier and it appears directly at the output for a trans impedance amplifier. Hence for a trans-impedance amplifier, the OTA noise amplification does not depend on the closed loop gain. Feedback resistor adds noise directly at the output. Increasing the feedback resistor increases its noise contribution but amplifies the signal itself. The output referred noise of the transimpedance amplifier is

$$v_{n,2}^2 = 4kTR + v_{n,\text{OTA2}}^2 \tag{4.31}$$

where $v_{n,\text{OTA2}}$ is the input-referred noise of OTA$_2$.

The first stage in OTA$_2$ is designed with a large overdrive voltage for linearity reasons. In the TIA the OTA gain and bandwidth are not affected by the closed loop gain except for loading, unlike for an inverting amplifier, which allows a 10 dB improvement in GBP and DC gain which yields a 28 dB improvement in the IMD for a closed loop gain of $n = 2$.

4.6.3   Anti-Alias Filter

The loop gain for a well compensated system rolls of at the rate of 20 dB/dec near the unity gain frequency. The unity gain frequency of the loop gain is approximately the 3 dB bandwidth of the closed loop system. The nonlinearity is suppressed by the loop gain in a feedback system. However since the loop gain starts to roll off well before the 3 dB bandwidth of the closed loop system, the linearity performance is degraded for signals near the 3 dB bandwidth. Hence the unity gain frequency is typically much larger than the signal bandwidth. This allows for signals near the band edge to have a higher linearity. However increasing the loop unity gain frequency increases the integrated noise thereby degrading the noise performance. The noise of the first order system with 3 dB bandwidth of $\omega_{3\,db}$ is given by

$$v_n^2 = \frac{kT\gamma(1 + \eta)}{g_m}\omega_{3\,dB} \tag{4.32}$$

where η is the excess noise factor of the OTA. Hence in this work, we mitigate this using a passive RC circuit to do a first-order anti-alias filter utilizing R_f in Fig. 4.12 and the sampling capacitors. The 3 dB bandwidth of the RC circuit is selected to be two times the signal bandwidth to have minimal residual attenuation at the band edge frequency while filtering an appropriate amount of noise. Since it is a passive circuit, it doesn't deteriorate the linearity of the circuit. This resistor also servers as the compensation of the loop as the resistor in series with sampling capacitor creates a left half-plane zero which increases the phase margin of the loop. It also reduces any effects of switching transients from the sampler onto OTA$_2$.

4.6.4   Sampler

The input signal to the sampler is half V_{dd} (1 V$_{pp-diff}$) and is sampled on a 4 pF capacitor at 100 MS/s to be used for passive amplification at a later stage. We designed an normal RC sampler. Two samplers copies are clocked in a ping-pong fashion to maintain a constant OTA load. This features using a time interleaved ADC onto the two sampling capacitors.

4.6.5 Passive Amplification

Passive amplification is employed to get a rail-to-rail output [26]. A passive amplifier introduces less noise (kT/C) compared to an active amplifier ($\eta kT/C$, where η is the excess noise factor of the active amplifier) and is also inherently highly linear. As for conventional passive amplifier, sampled voltage is simply doubled by series connection of two capacitors in Φ_2 high phase in Fig. 4.12. The differential output V_{outd} and common-mode V_{outcm} are calculated as

$$V_{outd} = 2(V_{in+} - V_{in-}) \tag{4.33}$$

$$V_{outcm} = \frac{C_2}{C_1 + C_2} V_{dd} \tag{4.34}$$

In this design C_1 and C_2 are set to the same capacitance of 500 fF, which corresponds to the common mode voltage of 500 mV. Further this passive amplifier provides common-mode rejection and sets the output common-mode voltage of the ADC at half V_{dd}. In addition, it achieves an area reduction of the sampling capacitor and requires no reference voltage.

This ADC gives a gain of 8 with 2 due to current mirror in current domain, 2 due to ratio of resistors in converting voltage to current and back to voltage and 2 due to passive amplifier.

4.7 Measurements

The proposed ADC driver (Fig. 4.16) is fabricated in a 65 nm CMOS general purpose technology. The area occupied by the driver is 0.0084 mm^2 out of which 74 % are sampling capacitors. An open-drain PMOS (IO device) source-degenerated differential pair is used to drive the output off-chip. Phase and amplitude matched 50 Ω baluns were used to interface with single-ended equipment. Sharp band pass filters were used to clean the harmonics generated by the signal generators. The measurement results are discussed below. The sampler in ADC is clocked with 100 MHz clock to sample signal up to bandwidth of 50 MHz.

Figure 4.17 shows the magnitude response of the ADC driver up to 100 MHz. Although the sampling frequency is 100 MHz, signal magnitudes up to 100 MHz is measured by its aliased components. The flat magnitude response with no peaking indicates a well compensated system. The closed loop gain is 15.66 dB at the Nyquist frequency of 50 MHz and 17.9 dB at 1 MHz. The 3 dB frequency is 57 MHz which is higher than the signal bandwidth (50 MHz) to reduce any attenuation in signal bandwidth.

Fig. 4.16 Micrograph of the
ADC driver

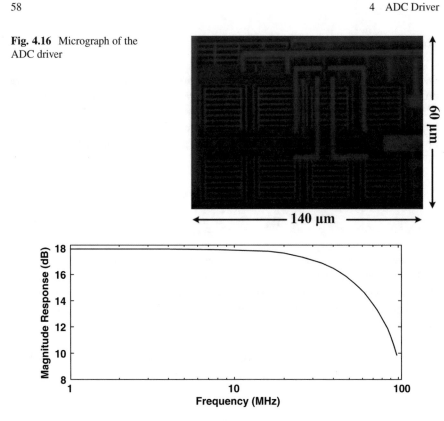

Fig. 4.17 Magnitude response of the ADC driver

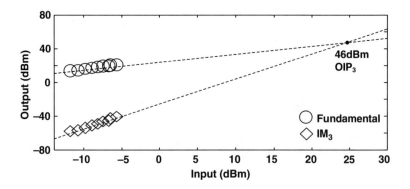

Fig. 4.18 Measured IIP$_3$ at 50 MHz using two tones with 1 MHz offset

The entire ADC driver consumes 2 mW of power from a 1 V supply. Figure 4.18 show the measured IIP$_3$ of the driver with input tones at 49 and 50 MHz. Third harmonic rises thrice as fast as the fundamental. The circles and diamonds indicate the measurement points of fundamental tone and third order intermodulated

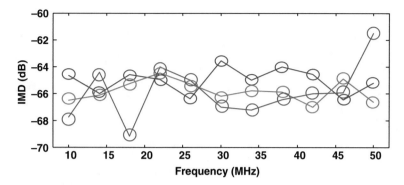

Fig. 4.19 Measured IMD for 2 V$_{pp\text{-}diff}$ output with 1 MHz tones separation. *Red*, *blue* and *green lines* indicate three different chips

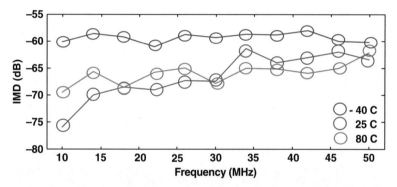

Fig. 4.20 Measured IMD for 2 V$_{pp\text{-}diff}$ output with 1 MHz tones separation at different temperatures

component. An OIP$_3$ of 46 dBm is achieved in the design and the output follows the OIP$_3$ extrapolation lines even for a rail-to-rail output swing. The measured IIP$_3$ is 24.82 dBm. The higher linearity even at rail-rail swing is due to the passive amplifier. Figure 4.19 shows the intermodulation distortion (IMD) across the input frequency for three different chips. The IMD for the full rail-to-rail swing is −67 dB and varies to −60 dB across ten chips. The temperature dependence of IMD is shown in Fig. 4.20. At the Nyquist frequency the IMD varies by only 4 dB over 80 °C due to the semi-constant current biasing of the inverters and the fact that all blocks in the design are multiples of the unit bias inverter. Figure 4.21 shows the intermodulation distortion at the Nyquist rate for rail to rail output swing across ten chips. Figure 4.22 shows a screen capture of a particular test result for two-tone inputs at 49 and 50 MHz.

In order to study the effect of transistor mismatch, a 100 point Monte Carlo simulation is done on ADC driver as shown in Fig. 4.23. The mean simulated value of the intermodulation distortion is obtained as 70 dB with standard deviation of 6.394 dB.

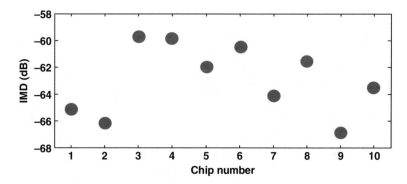

Fig. 4.21 Measured IMD with tones at 50 MHz separated by 1 MHz across chips

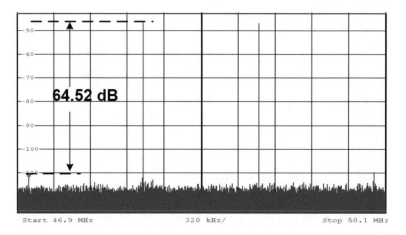

Fig. 4.22 Measured IMD with tones at 50 MHz separated by 1 MHz

Figure 4.24 show a screen capture of the noise measurement up to 50 MHz for the 100 MS/s output. We get a uniform noise floor because of aliasing of out of band noise. The buffer used in the noise measurement has an attenuation of 7 dB and hence the actual noise floor is at -124 dBm. The total integrated output noise is 1 mV$_{rms}$ resulting in an SNR of about 60 dB. Although this work is about an ADC driver with a voltage gain of 8 and a sampler, this design can be thought of as a first-order AAF for comparison purposes and hence we use the filter FOM [28].

Table 4.1 summarizes our design and compares it with other published state-of-the-art filters with cutoff frequencies in this range. The noise filtering by the series resistor and its ability to compensate the system lead to lower noise performance and lower power. For a fair comparison, we have divided the area by the order of the filter. We get the best IMD (-65 dB) at full rail-to-rail swing sampled at 100 MS/s. The output follows the OIP$_3$ extrapolation lines even at rail to rail swing and the OIP$_3$ measured is 46 dBm. Although power of a system increases with gain, we get the state of art figure of merit and power when compared to a filter of gain 1.

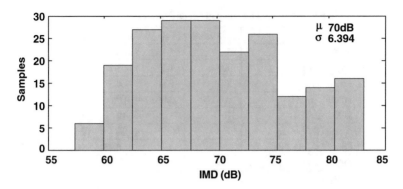

Fig. 4.23 Simulated Monte Carlo analysis on IMD

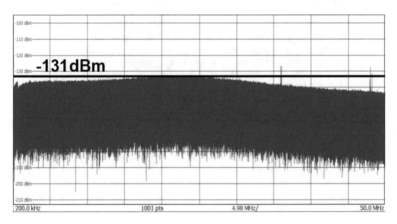

Fig. 4.24 Screen capture of the noise measurement

The driver can take a small signal swing directly from the AGC/ RF front end and amplify to rail to rail for ADC while still maintaining the SNR and SFDR for 10 bit accuracy.

4.8 Conclusion

The proposed architecture is significant because it can bridge the gap between the limited swing provided by circuits preceding ADCs and the power and noise advantages provided to ADCs in scaled technologies by full scale input swings. This work achieves the best output swing to V_{DD} ratio at lower power supplies with a linearity (IMD) of 65 dB at Nyquist rate with comparable figure of merit (35 aJ) with a DC gain of 8. A semi-constant current biasing scheme reduces the PVT variation. The intermodulation distortion variation across ten chips is 6 and 4 dB across a temperature variation of 120 °C.

Table 4.1 Summary and comparison with prior art

	This work	[17]	[28]	[25]
Tech (nm)	65	65	180	90
Supply (V)	1.0	1.2	1.8	0.9
Power (mW)	2.0[a]	1.56	4.1	19.1
Area/N (mm^2)	0.0084[a]	0.0057	0.065	0.073
BW (MHz)	57	58.7	10	30
Noise (nV/$\sqrt{\text{Hz}}$)	17	80.7	20	26.0
Output swing ($V_{pp\text{-}diff}$)	2	0.4	0.75	0.33
O/P ratio ($V_{pp\text{-}diff}/V_{DD}$)	2.0	0.33	0.42	0.37
OIP$_3$ near f_c (dBm)	46	33	41.5	34.3[b]
DC gain	8	1	1	1
SFDR[b] near f_c (dB)	60	55.8	60.3	65.5[c]
FOM (aJ)	35	23.1	42.0	44.9

[a]Includes sampling capacitor area, drive and clocking power
[b] SFDR (dB) $= \frac{2}{3}\left[\text{OIP}_3 \text{ (dBm)} - \text{Gain (dB)} - P_{\text{noise}} \text{ (dBm)}\right]$
[c]Estimated SFDR: DR=65.5 dB, low freq. THD=65.2 dB

Chapter 5
Current Mirror Based Filter

Analog baseband circuits, including filters and programmable gain amplifiers (PGA), are indispensable in wireless sensors and communication systems. Portable devices continue to drive the need for circuits that achieve low power without sacrificing linearity. These analog baseband filters typically consume tens of mW and have a considerable impact on the total power consumption of a transceiver. Wireless receivers today demand low-noise and linear baseband channel select filters to accommodate the large dynamic range of the input signals. Power dissipation and die area of these filters must also be as small as possible. Analog frond ends are usually integrated with digital function blocks in recent mixed-signal Soc applications including data communication and image processing systems. Low power and small area ADCs with 8–10b accuracy operating several tens of MS/s are considered important building blocks. However each ADC has to he preceded by a front end anti-alias filter with cut off frequency in MHz range. There are two major concerns when we design a wideband LPF. One is the selection of filter LCR prototype, that is, Butterworth, Chebychev, Bessel etc. The other is the selection between G_m-C and active-RC.

Active filters are implemented using G_m-C, active-RC, or G_m-OTA-C techniques [29]. G_m-C filters are capable of high-speed operation due to their open-loop nature [30]. Open-loop operation also means that the G_m-C filter linearity is limited by the transistor's inherent non-linearity. Increasing the overdrive voltage improves the linearity at the expense of higher power and higher noise. Any attempt to linearize the G_m results in a higher noise. Another way to improve G_m-C filter linearity is through the use of source follower based G_m-C filters [31] to exploit the increase in linearity with reduced overdrive voltage, resulting in high linearity and lower power consumption. The major drawback, however, is the limited output swing. Active-RC filters use OTA-based integrators, resulting in a linearity that is only limited by the swing at the virtual ground node of the amplifier [28]. Hence, their linearity and noise performance is usually better than that of G_m-C

© Springer International Publishing AG 2017
R.K. Palani, R. Harjani, *Inverter-Based Circuit Design Techniques for Low Supply Voltages*, Analog Circuits and Signal Processing,
DOI 10.1007/978-3-319-46628-6_5

filters. However, bandwidth requirements of the amplifier result in a higher power consumption. Amplifier efficiency can be improved through the use of push-pull buffers, leading to an effective increase in the filter bandwidth [32] at the expense of output swing.

Active RC filters are realized typically using an active inductor obtained by cascading integrators. The quality factor of the integrators determine the maximum quality factor that can be obtained in these filters. Active RC filter can be derived from an low pass RC circuit as discussed below. Figure 5.1 shows the passive RC circuit and its feedback model. The resistor R converts the difference between the input and output voltage to current. The current gets integrated onto the capacitor. The passive RC circuit can be assumed as an integrator with unity gain feedback around it. The integrator can be realized by breaking the feedback. One way of breaking the feedback is to make the current created by resistor independent of the output voltage. The capacitor node can be held at virtual node by sinking the current V_{in}/R through the capacitor from an OTA as shown in Fig. 5.2. Now the resistor converts only the input voltage rather than the difference between input and output voltage to current. This breaks the feedback loop to realize an active RC integrator.

The pole of passive circuit lies in the left half plane as they are lossy. The active RC integrator moves the pole of a passive RC filter from left half plane to origin. Active RC filters use cascade of active integrators to realize the transfer function. Hence active RC filters moves the poles of integrator from origin again to left half complex plane to realize the filter transfer function. In this work, Butterworth filter is designed by moving the poles of the passive low pass filter directly to the complex plane rather than moving the poles to origin and then to complex plane. The crosses in Fig. 5.3 indicate the location of the poles. The black crosses are the original poles due to an passive RC circuit. In conventional integrator based architectures, the poles move from left half plane to origin (red crosses) and then to appropriate

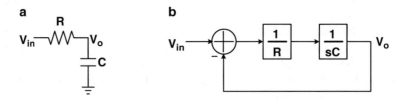

Fig. 5.1 Passive RC low pass circuit (**a**) and its feedback model (**b**)

Fig. 5.2 Active RC integrator

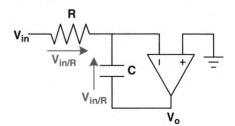

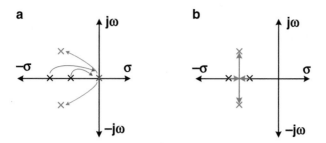

Fig. 5.3 Poles in an active RC filter. (**a**) Conventional biquad poles; (**b**) proposed biquad poles

pole location as dictated by the filter. However in proposed architecture poles move directly to the complex plane. This gives an advantage of power and noise due to reduced low impedance nodes over conventional active RC filters as explained in filter architecture

We present a new filter design that relaxes the power-linearity trade-off. Unlike conventional integrators, in the proposed design the unity-gain frequency (UGF) of the overall integrator coincides roughly with the UGF of the amplifier. Moreover, signal processing is performed in the current domain throughout the whole signal chain, reducing the number of low impedance nodes and power supply limitations. Also, no additional compensation capacitors are needed since the filter capacitors themselves compensate the amplifier. As will be shown, this results in both reduced area and power consumption.

5.1 Integrator Design

An active-RC or G_m-C integrator, as shown in Figs. 5.4a and 5.5a, form the core of integrated continuous-time active filters. Due to the high quality of integrated resistor and capacitors available, the performance of these topologies are primarily limited by the linearity, gain, and bandwidth of the integrator. Here, we introduce a new design and compare it to conventional designs. The OTA is realized as a simple five-transistor differential pair. This is followed by an analysis of the gain and bandwidth limitation effects upon them and the advantages provided by the proposed design.

In the case of an active-RC filter, we shall assume that the amplifiers can be broken up into a high-gain stage followed by a class AB driver (inverter) for best power efficiency as shown in (a). The conventional active-RC structure, shown in Fig. 5.4a, is modified as shown in (b) to form the proposed inverter-based integrator design. The loading of the integration capacitor is removed from the feedback loop, easing the bandwidth requirement of the amplifier. The active elements now effectively function as a current mirror with scaling factor k determined by inverter sizing. The inverters have a high g_m/I_d efficiency, wide swing, and scale very well with

Fig. 5.4 A conventional
active-RC integrator (**a**) and
the proposed integrator (**b**)

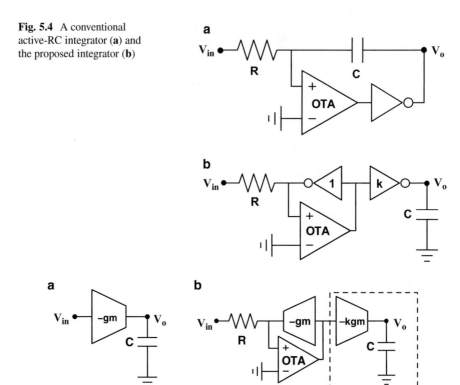

Fig. 5.5 A conventional G_m-C integrator (**a**) and a functional diagram of the proposed design
which linearizes its G_m-C output section (**b**)

technology. The additional degree-of-freedom introduced by adding the current gain
can be used to control the root locus of the proposed biquad in Sect. 5.2.1.

The proposed integrator can also be compared to a G_m-C integrator by redrawing
the circuit as in Fig. 5.5b. The nonlinearity of a G_m-C integrator is primarily
determined by the nonlinearity in the transconductance and of the finite output
impedance. As in G_m-C integrators, increasing the channel length and cascoding
the output reduces the non-linearity effects due to the finite output impedance.
In this case, the overall nonlinearity is dominated by the nonlinearity in the
transconductance.

In the proposed design, shown in Fig. 5.5b, the current supplied by the feedback
transconductance is always equal to the current introduced by the resistor. Any
nonlinearity in the feedback transconductance is reflected by a change in the output
voltage of the high-gain stage. As the feedback transconductance ($-g_m$) and the
output transconductance ($-kg_m$) are matched, any nonlinearity introduced by the
transconductor is canceled. Non-linearity in the current mirror is due to g_m and
g_{ds} non-linearity of the transistors. The OTA in negative feedback adjusts the gate
voltage of gm cell (gm) to absorb the input current. The input current is created by

using a passive resistor and hence it is linear. If the transconductance of gm cell (gm) is assumed to have an nonlinear function $f(x)$, then the gate voltage of the gm cell (gm) can be expressed as

$$V_g = f^{-1}\left(\frac{V_{in}}{R}\right)$$
(5.1)

Since the gm cells gm and kgm share the same gate voltage, the output current is the amplified version of the input current and is given by

$$I_o = kf(V_g) = kf\left(f^{-1}\left(\frac{V_{in}}{R}\right)\right) = k\frac{V_{in}}{R}$$
(5.2)

Our design relies on the matching of the transistors. For this reason the inverters use a common-centriod layout to reduce systematic variations. As can be seen intuitively, as in G_m-C filters, any extra load capacitance (ADC sampling capacitor) can be effectively absorbed into this filter capacitor. This enables to sample the filtered signal onto the ADC sampling capacitor without additionally loading the amplifier.

5.1.1 Non-Linearity Cancellation

This design primarily relies on the non-linearity cancellation (Fig. 5.6) of a current mirror. Here resistor R converts voltage to current and it is absorbed by the inverter. The OTA produces a voltage at the gate of inverter which is required to absorb that current. If the trans conductance of the inverter is assumed to have the transfer function f(V) then the gate voltage required to absorb the current V_{in}/R is given as $f^{-1}(V_{in}/R)$. Since this voltage is given to an identical inverter the output current of that inverter will be $f(f^{-1}(V_{in}/R)) = (V_{in}/R)$ canceling the non linearity of the inverters. If the transistors follow square law, OTA forces the voltage V_{int} at the gate of the inverter.

$$V_{int} = \sqrt{\left(\frac{2V_{in}}{R\beta}\right)} + V_T$$
(5.3)

where $\beta = \mu_n C_{ox} W/L$ and V_T is the threshold voltage. Since this is the gate voltage of another identical inverter output current is given by

$$I_o = \frac{\beta}{2}(V_{int} - V_T)^2 = \frac{V_{in}}{R}$$
(5.4)

5.1.1.1   Effect of Mismatch in Inverters

The Inverters I_1 and I_2 can have mismatch between them, primarily due to mismatch of the threshold voltage and aspect ratio in transistors resulting in an offset. If we model the offset in the inverters as V_{off} as shown in Fig. 5.6, then

$$V_{int} = f^{-1}(\frac{V_{in}}{R}) \tag{5.5}$$

$$I_o = f(V_{int} + V_{off}) \approx f(V_{int}) + V_{off}f'(V_{int}) \tag{5.6}$$

$$I_o = f(f^{-1}(\frac{V_{in}}{R})) + V_{off}f'(V_{int}) \tag{5.7}$$

$$I_o = \frac{V_{in}}{R} + g_m V_{off} \tag{5.8}$$

For offset voltages less than the input voltage we can approximate the function f using Taylor series as shown in Eq. (5.8). This will result in an offset current. However if the V_{off} is comparable with the input voltage V_{int}, g_m in Eq. (5.8) will be nonlinear resulting in the incomplete non linear cancellation. Figure 5.7 shows the 100 point Monte Carlo simulation on the current mirror. The output current is taken at the low impedance node to isolate the gm non linearity. The simulated mean intermodulation distortion is 105.6 dB and the standard deviation is 8.2 dB. For best nonlinear cancellation the trans conductance of the inverters should be as small as possible which in turn translates to lower aspect ratio and lower current. The random mismatch can be reduced by increasing both the width and length of the transistor proportionally. The lower transconductance on the current mirror also results in lower noise.

Fig. 5.6 Non-linear cancellation in proposed integrator

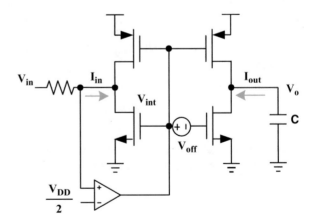

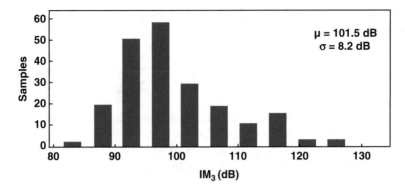

Fig. 5.7 Monte Carlo simulation on the current mirror

5.1.1.2 Comparison with Conventional Gm-C Integrator

In an convention Gm-C integrator the non-linearity is primarily governed by the transconductance(g_m) linearity and finite output impedance (g_{ds}) linearity. Higher the transconductance of higher bias current results in lower g_m nonlinearity and higher the output impedance results in lower output impedance non-linearity. However higher bias current to reduce the g_m non-linearity reduces the output impedance of the transistor thereby increasing the g_{ds} non-linearity. Any g_m linearization technique results in higher noise. Hence optimal bias current is at the point where the non-linearity due to g_m and g_{ds} becomes equal if it is not limited by the noise.

In this technique since we cancel the non-linearity of the trans conductance, the g_m non-linearity is reduced resulting in the non-linearity only due to g_{ds}. Since lower g_m is appreciated for lower noise and lower mismatch, the transconductances are biased with lower biasing current and lower aspect ratio resulting in higher output impedances. This translates to lower g_{ds} non-linearity and higher Quality factor in the filter.

5.1.2 Bandwidth Limitation Effects

In the integrator design of , the amplifier should be compensated and its transfer function can be approximated by $H(s) = A_0/(1 + s/\omega_p)$, where A_0 is the DC gain and ω_p is the 3 dB frequency of the compensated two-stage system. We study the effect of finite gain and the finite unity gain bandwidth separately. The finite gain of the amplifier results in a gain error in both circuits. In order to understand the effect of finite unity gain bandwidth of the amplifier upon the integrator, we assume the DC gain to be infinite and its unity gain bandwidth as ω_u so the transfer function of the amplifier can be approximated as ω_u/s.

The transfer function of an active based integrator with bandwidth limitation is thus given by below.

$$H_{rc}(s) = \frac{-1}{sRC\left(1 + \frac{1}{RC\omega_u}\right)\left(1 + \frac{s}{(\omega_u + 1/RC)}\right)} \qquad (5.9)$$

The finite ω_u primarily has two effects on the integrator. First, it modifies the unity gain frequency of the overall integrator. Second, it introduces additional phase delay. The result is that the UGB (ω_u) of the amplifier has to be much greater than the UGB of the integrator so the effect upon the integrator is negligible. For example, if ω_u is four times $1/RC$, then the UGB of the integrator is reduced by 20 %.

In the inverter-based design, the finite ω_u does not alter the unity gain frequency of the integrator. However, the additional phase delay due to the finite ω_u still remains. The transfer function of the proposed inverter-based integrator can be derived as shown in below.

$$H_{pr}(s) = \frac{-1}{sRC\left(1 + \frac{s}{\omega_u}\right)} \qquad (5.10)$$

When we use this integrator in a biquad, the delay increases the Quality factor of the filter.

5.1.3 Gain Limitation Effects

The effect of finite DC gain of infinite bandwidth amplifier on active-RC and proposed integrator is given in Eq. (5.11). The frequency independent scaling factor due to finite gain is $A_f = A_0/(A_0 + 1)$ and r_0 is the output impedance of the G_m cell.

$$H_{rc} = -\frac{A_f}{\frac{1}{A_0+1} + sRC} \qquad H_{pr} = -\frac{A_f}{\frac{R}{r_0} + sRC} \qquad (5.11)$$

As seen in H_{rc}, the DC gain of the active-RC based integrator is primarily dependent on the gain of the amplifier. For the proposed integrator, H_{pr} shows that, like a G_m-C integrator, the DC gain is primarily limited by the output impedance of the transconductor.

5.1.4   Noise Analysis

The inverter, resistor and the OTA contributes noise. The output noise spectral densities of resistor, inverter and OTA is given by

$$I_R^2 = \frac{4kT}{R} \tag{5.12}$$

$$I_I^2 = 4kT\gamma(g_{mp} + g_{mn}) \tag{5.13}$$

$$I_{OTA}^2 = 4kT\gamma g_m(1 + \eta) \tag{5.14}$$

where g_{mp}, g_{mn}, g_m are transconductances of the pmos and nmos transistors in the inverter and OTA, η is the excess noise factor of the OTA. The input referred noise of the integrator is given by

$$v_n^2 = 4kTR\left(1 + \gamma\left(g_{mp} + g_{mn}\right)\left(1 + \frac{1}{m}\right)R + \frac{\gamma(1 + \eta)}{g_mR}\right) \tag{5.15}$$

The resistor noise and the input inverter noise directly adds at the input. Output inverter noise is divided by the current mirror ratio when referred to the input. OTA noise is divided by g_mR when referred to the input. For lower noise performance the inverter trans conductance should be as small as possible. This is obtained by using lower aspect ratio for transistors in inverters. However the OTA should have the largest trans conductance for lower input referred noise (Fig. 5.8).

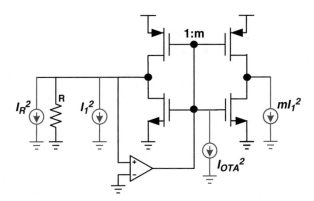

Fig. 5.8 Noise sources in the proposed integrator

5.2 Filter Design

Unlike active-RC or G_m-C integrators, the proposed integrator cannot be directly cascaded due to finite low input and high output impedance associated with them. Cascading the proposed integrators results in a low pass filter. It is shown in that a biquad can be designed using cascaded low pass filters and a gain element in feedback. The negative feedback pushes the real poles to the complex plane resulting in required Butterworth response. Unlike traditional active-RC filters, this filter is designed in current mode. In Fig. 5.9, all the inverters are shown as transconductance cells as we will be operating them in their active region. The common mode feedback circuit is not shown for the sake of clarity.

5.2.1 Current-Domain Biquad

The low pass filter current mode biquad is shown in Fig. 5.9. The resistor R_1 with OTA virtual node converts the input voltage into current. The biquad consists of two real poles placed in negative feedback with gain block. Resistors R_2, R_3 and capacitors C_2, C_3 provides two real poles. The gain is obtained by the current mirror ratio k_2, k_3. The gain in negative feedback moves the poles from real axis to desired filter pole locations on the complex plane. The second low pass filter in the biquad has a positive gain. However, in a differential architecture we can swap the wires for positive gain for the output inverter G_m. The DC gain, center frequency ω_o, and quality factor (Q) for this biquad are as follows:

$$A_{DC} = -\frac{R_3}{R_1}\left(\frac{k_2 k_3}{k_2 k_3 + 1}\right) \qquad \omega_o = \sqrt{\frac{k_2 k_3 + 1}{R_2 C_2 R_3 C_3}}$$

$$Q = \omega_o \cdot \frac{R_2 C_2 R_3 C_3}{R_2 C_2 + R_3 C_3} \tag{5.16}$$

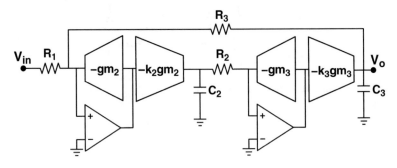

Fig. 5.9 Current mode low pass filter based biquad

The poles of this biquad are given by Eq. (5.17) where $U = R_2C_2 + R_3C_3$ and $V = R_2C_2R_3C_3$.

$$\omega_{p1,p2} = \frac{-U \pm \sqrt{U^2 - 4(k_2k_3 + 1)V}}{2V} \tag{5.17}$$

DC gain is primarily decided by the resistor R_1 which converts the voltage to current and the resistor (R_3) which converts the current back to voltage and also on the current mirror ratio (k_2, k_3). In order to have unity DC gain the resistor R_3 is chosen to be $R_1(1 + k_2k_3)/k_2k_3$. The cut off frequency is independent of the resistor R_1 which converts voltage to current since the biquad is in current mode and only R_2, R_3, C_2, C_3 steer the current in the biquad. If we have the products R_2C_2 and R_3C_3 to be equal then we have the quality factor independent of the filter components to be

$$Q = \frac{\sqrt{1 + k_2k_3}}{2} \tag{5.18}$$

This allows for the constant Q at different cut off frequencies.

If the current gain product k_2k_3 is greater than $(R_3C_3 - R_2C_2)^2/(4R_2C_2R_3C_3)$, the poles split into a complex conjugate pair. These poles can be appropriately placed in the complex plane by choosing the current gain. This condition is easily met for practical design values (i.e. k_2 and $k_3 > 1$, and RC time-constants not overly distant from one another). The quality factor can be increased by choosing a higher current gain (k_2 and k_3) unlike in an active-RC or G_m-C filter where it is limited by the DC gain of the amplifier or the finite output impedance of the transconductor, respectively.

5.2.2 Effect of OTA Nonidealities on Biquad

The OTA is used in negative feedback to create low impedance nodes and is realized using a five transistor differential pair. The OTA is assumed to be one pole system with dc gain of A_{OTA} and 3 dB bandwidth of ω_p as shown below

$$\frac{V_{out,OTA}}{V_{in,OTA}} = \frac{A_{OTA}}{1 + \frac{s}{\omega_p}} \tag{5.19}$$

The transfer function of the biquad can be approximated as

$$\frac{V_o}{V_{in}} = \frac{A_{DC}\omega_o^2}{s^2 + s\frac{\omega_o}{Q} + \omega_o^2} \frac{1}{1 + \left(\frac{s^2}{\omega_u} + s\frac{2}{\omega_u}\right)\left(\frac{s^2 + s\frac{\omega_o}{Q} + \frac{\omega_o^2}{1 + k_2K_3}}{s^2 + s\frac{\omega_o}{Q} + \omega_o^2}\right)} \tag{5.20}$$

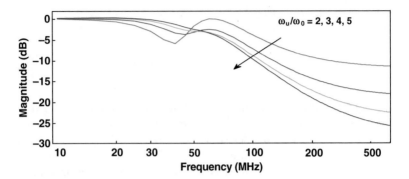

Fig. 5.10 Effect of OTA non-idealities on biquad

Here ω_u is the unity gain frequency of the negative feedback loop. The first part of Eq. (5.20) is the ideal biquad transfer function corresponding to an ideal operational amplifier. The finite gain and bandwidth of the OTA creates a parasitic biquad as shown in Eq. (5.20) whose quality factor and the natural frequencies are 0.5 and ω_o. The biquad is simulated for various unity gain frequency of the loop as shown in Fig. 5.10. The effect of parasitic biquad on the magnitude response of the biquad is minimal for $\omega_u > \omega_o$.

5.2.3 Butterworth Filter Design

For the prototype design, we cascade a first order system and this biquad in the current domain to obtain a third-order Butterworth filter as shown in Fig. 5.11. The transfer function of the filter can be shown to be

$$\frac{V_o}{V_{in}} = -\frac{R_3}{R_0}\left(\frac{k_2 k_3}{k_2 k_3 + 1}\right)\left(\frac{1}{1 + sR_1 C_1}\right) \times \left[\frac{1}{s^2\left(\frac{R_2 C_2 R_3 C_3}{k_2 k_3 + 1}\right) + s\left(\frac{R_2 C_2 + R_3 C_3}{k_2 k_3 + 1}\right) + 1}\right]$$
$$(5.21)$$

The swing on the capacitors C_1 and C_2 is kept small by choosing larger value of capacitor and smaller resistor. This helps in reducing the non-linearity due to finite output impedance and also helps in compensating the negative feedback loops. However the swing on the capacitor C_3 is kept same at that of the input at DC to ensure the DC gain of one. The DC gain, cut off frequency ω_o and quality factor of the filter is given by Eq. (5.24).

$$A_{DC\,filter} = \frac{R_3}{R_0}\frac{k_2 k_3}{1 + k_2 k_3}$$
$$(5.22)$$

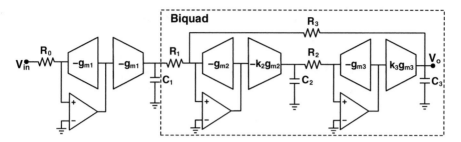

Fig. 5.11 Schematic of the third order filter using the proposed integrator and current-mode biquad

$$\omega_o = \sqrt{\frac{k_2 k_3 + 1}{R_2 C_2 R_3 C_3}} \tag{5.23}$$

$$Q = \omega_o \cdot \frac{R_2 C_2 R_3 C_3}{R_2 C_2 + R_3 C_3} \tag{5.24}$$

The DC gain primarily depends on the ratio of resistors which is used to convert voltage and current and the ratio of gm cells and is set to one. The cut off frequency is set to 50 MHz and the quality factor of one as required by third order Butterworth filter.

5.2.4   Compensation of the Amplifiers

This filter has three amplifiers in negative feedback. Each amplifier is a two stage Class AB design. This system has to be compensated for stable operation. The first stage amplifier has a resistive load and the second and third stage amplifier have a series resistor and capacitor combination load. Therefore, different design principles apply for these two designs. The first amplifier is compensated by using conventional Miller compensation.

The second and third stage amplifiers pump current into the series resistor and capacitor combination as shown in Fig. 5.12. This generates a left half plane zero. If we make the input resistor smaller than the output impedance of inverter and the filter capacitor to be greater than the parasitics, the amplifier can be compensated with this zero. The loop is broken at the OTA input to study the loop dynamics as shown in Fig. 5.13. The transconductance gm_1 and gm_2 represents the OTA and the inverter respectively. C_1 and C_2 represents the parasitics of the devices, g_{ds1} and g_{ds2} represents the output impedance of the OTA and the inverter, R and C represents the filter resistor and capacitor. The loop gain is given by

$$\frac{V_{out}}{V_T} = A_{DC} \left(\frac{1 + sRC}{s^2 \frac{RCC_2}{g_{ds2}} + s\left(\frac{C_2}{g_{ds2}} + \frac{C}{g_{ds2}} + RC\right) + 1} \right) \left(\frac{1}{1 + s\frac{C_1}{g_{ds1}}} \right) \tag{5.25}$$

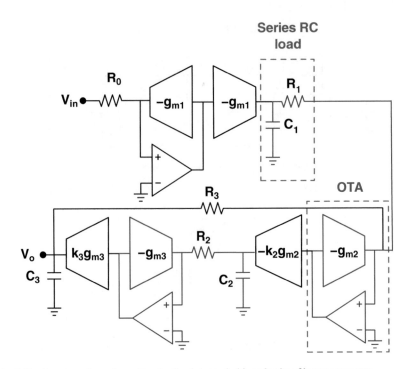

Fig. 5.12 Compensation of negative feedback loops in biquad using filter components

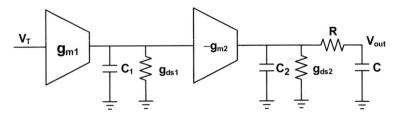

Fig. 5.13 Schematic of loop gain of one stage in biquad

where

$$A_{DC} = \frac{g_{m1}g_{m2}}{g_{ds1}g_{ds2}} \qquad (5.26)$$

The system has three real poles and one zero. There is one real pole due to the OTA parasitics at g_{ds1}/C_1 and two poles due to the inverter parasitics and the filter components. Filter resistor and the capacitor creates both pole and a zero. If we assume that the filter resistor R is less than the output impedance of the transistor and the filter capacitor is much greater than the transistor parasitics and the poles at the inverter output are far apart, we can obtain the pole and zero at

$$P1 = -\frac{g_{ds2}}{C} \quad P2 = -\frac{g_{ds1}}{C_1} \quad P3 = -\frac{1}{RC_2} \qquad (5.27)$$

$$Z = -\frac{1}{RC} \qquad (5.28)$$

Since filter resistor and transistor parasitics is less than the output impedance and parasitics of transistor, pole P3 is at the higher frequency. The pole P1 is at the lower frequency than P2 and zero is in between them. Both the Pole P1 and Z moves with the filter capacitor and hence the unity gain frequency of the loop is independent of the filter capacitor to first order. Unity gain frequency (ω_u) and the phase margin is approximated as

$$\omega_u = \frac{g_{m1}g_{m2}R}{C_1} \qquad (5.29)$$

$$PM = 2\pi - tan^{-1}\left(\frac{g_{m1}g_{m2}R^2C/C_1}{1 + Rg_{ds1}C/C_1}\right) \qquad (5.30)$$

Filter capacitor and resistor along with inverter parasitics produces two pole and a zero, one pole and zero close by and other pole at frequency higher than the unity gain frequency (UGB) of the loop gain. OTA produces pole P2 which is within the UGB of the loop gain. In effective we have two poles and one zero within the UGB. For the same cut off frequency 1/RC, increasing the capacitor pushes the pole P1 to lower frequency and P3 to higher frequency. Further separation of poles P1 and P2 is increased with increasing C. Since the zero location is unaltered, increasing capacitor C increases the phase margin of the system thereby compensating the system.

Since there is no addition compensation technique used and the load (filter resistor and capacitor) itself compensates the system, the unity gain bandwidth obtained can be higher than the traditional compensation techniques for given power. Unlike integrator based active RC filters, lower filter resistor does not affect the DC gain of the negative feedback network as it is in series with the capacitor.

Monte Carlo simulation is done to see the effect of mismatch on the unity gain frequency and phase margin as shown in Fig. 5.14. The mean unity gain frequency and phase margin if 959.6 MHz and 60.28° with standard deviation of 130 MHz and 7.78°. We simulated the loop parameters at process corners to see the effect of global variations as shown in Fig. 5.15. Unity gain frequency is at least 4x greater than the cut of frequency over all the corners except slow. This is because at slow-slow corner the OTA transistors enter triode region. However the phase margin is greater than 45° over the entire space of process corners. The inverters in this design are biased using traditional replica metastable point biasing. Hence the unity gain frequency varies from 0.2 to 1.2 GHz over the process corners.

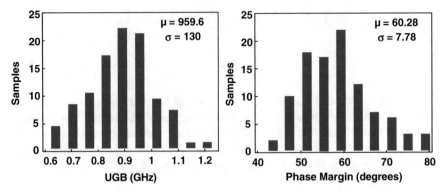

Fig. 5.14 Monte Carlo simulation on a negative feedback loop in biquad

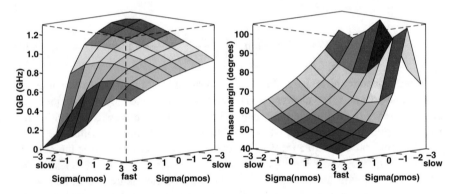

Fig. 5.15 Corner simulation on a negative feedback loop in biquad

5.2.5 Noise Comparison with Active RC Integrator Filter

The proposed filter has fewer low impedance nodes and hence has noise advantage over an active RC integrator based filter. Noise is contributed by resistors and transistors. All the resistors for analysis are assumed to be of equal value (5k) and capacitors are chosen to obtain a cut off frequency of 50 MHz ($k_2k_3 = 3$). Further it is assumed that the noise contributed of the active element is same as that of the passive elements for simplicity.

5.2.5.1 Active RC Filter

Figure 5.16 shows the active RC integrator based third order Butterworth filter. The noise of the resistors are represented by current source and OTA by its input referred voltage noise. The noise sources I_{R1}, I_{R2}, V_{OTA1} adds directly to the input signal. The transfer function for these noise sources to the output is given as

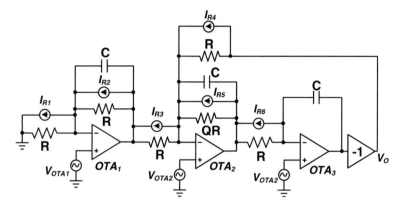

Fig. 5.16 Noise sources in the active RC integrator based filter

$$H(s) = \frac{\omega_0^2}{s^2 + s\frac{\omega_0}{Q} + \omega_0^2} \tag{5.31}$$

$$\frac{v_o^2}{i_{R1}^2} = \left|\frac{1}{1 + sRC}H(s)\right|^2 R^2 \tag{5.32}$$

$$\frac{v_o^2}{i_{R2}^2} = \left|\frac{1}{1 + sRC}H(s)\right|^2 R^2 \tag{5.33}$$

$$\frac{v_o^2}{V_{OTA1}^2} = \left|\frac{2 + sRC}{1 + sRC}H(s)\right|^2 \tag{5.34}$$

The OTA_1 in Fig. 5.16 generates a low impedance node at the output due to negative feedback. Entire noise current I_{R3} flows into this low impedance node creating the noise voltage. Similarly the entire noise current I_{R4}, I_{R5} directly create noise voltage at the output. The input referred OTA noise V_{OTA2} is shaped by the biquad transfer function H(s) and appears at the output.

$$\frac{v_o^2}{i_{R3}^2} = |H(s)|^2 R^2 \tag{5.35}$$

$$\frac{v_o^2}{i_{R4}^2} = |H(s)|^2 R^2 \tag{5.36}$$

$$\frac{v_o^2}{i_{R5}^2} = |H(s)|^2 R^2 \tag{5.37}$$

$$\frac{v_o^2}{V_{OTA2}^2} = |1 + H(s)|^2 \tag{5.38}$$

Similarly OTA_2 creates a low impedance node at the output. The entire noise I_{R6} and V_{OTA3} are shaped by the transfer function and appears at the output.

$$\frac{v_o^2}{i_{R6}^2} = \left| \frac{sCR}{1 + sCR + s^2R^2C^2} \right|^2 R^2 \tag{5.39}$$

$$\frac{v_o^2}{V_{OTA3}^2} = \left| \frac{(1 + sCR)^2}{1 + sCR + s^2R^2C^2} \right|^2 \tag{5.40}$$

All the components adds noise directly at the output except I_{R6} which is band pass filtered.

5.2.5.2   Proposed Filter

The proposed filter has the cascade of low pass filter and a biquad. The active element (OTA and inverter) noise is assumed equal to the passive and the noise is represented by the input current source. Since input is the voltage source, like active RC, the noise of resistor and active IR_0 and I_{io} adds directly to the signal. Input referred current noise spectral density of the resistor and current mirror is given as $4kT/R$ and $4kTg_m(1 + 1/m)$ where m is the current mirror ratio. Input referred voltage noise spectral density is given as $4kT(1 + \eta)/g_m$. The transfer function of these noise source to the output is given by

$$H(s) = \frac{\omega_0^2}{s^2 + s\frac{\omega_0}{Q} + \omega_0^2} \tag{5.41}$$

$$\frac{v_o^2}{i_{R0}^2} = \left| \frac{1}{1 + sRC} H(s) \right|^2 R_0^2 \tag{5.42}$$

$$\frac{v_o^2}{i_{i0}^2} = \left| \frac{1}{1 + sRC} H(s) \right|^2 R_0^2 \tag{5.43}$$

However the output of the first order filter is a high impedance node. Hence the noise current I_{R1} is high pass filtered by R_1 and C_1 and appears at the output. The transfer function for each noise source to the output is given by

$$\frac{v_o^2}{i_{R1}^2} = \left| \left(\frac{sR_1C_1}{1 + sR_1C_1} \right) H(s) \right|^2 A_{DC}^2 R_1^2 \tag{5.44}$$

$$\frac{v_o^2}{i_{R2}^2} = |sR_2C_2H(s)|^2 A_{DC}^2 R_1^2 \tag{5.45}$$

$$\frac{v_o^2}{i_{R3}^2} = \left| \left(1 + s\frac{R_1C_1}{1+k_2k_3} \right) H(s) \right|^2 R_1^2 \tag{5.46}$$

$$\frac{v_o^2}{i_{i1}^2} = |H(s)|^2 A_{DC}^2 R_1^2 \tag{5.47}$$

$$\frac{v_o^2}{i_{i2}^2} = \left| \frac{1}{k_2} (1 + sR_2C_2) H(s) \right|^2 A_{DC}^2 R_1^2 \tag{5.48}$$

Current mirror noise (I_1, I_2) and the feedback resistor R_3 noise adds directly at the output. However the OTA and the resistor R_1 and R_2 noise are shaped by the filter. Current mirror noise can be reduced by lowering its trans conductance which translates to lower power (Fig. 5.17). Only three out of six noise sources contribute noise to the output. unlike traditional active RC low pass filter design where all the sources adds noise directly at the output. Active RC integrator based design and the proposed low pass filter based design are simulated assuming all the active elements noise are equal to the passive elements noise to study the noise performance. Figure 5.18 shows the comparison for noise power spectral density at

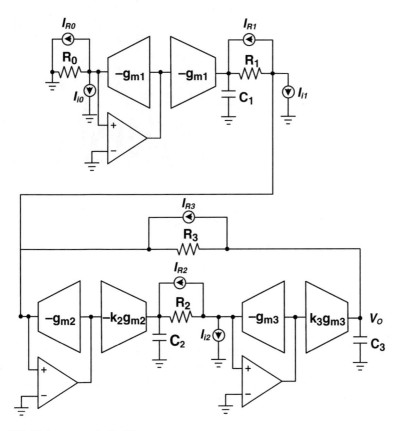

Fig. 5.17 Noise sources in the filter

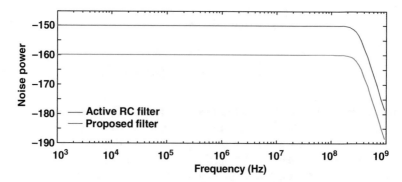

Fig. 5.18 Noise comparison between active RC and proposed filter

Fig. 5.19 Chip micrograph
and the test board used for
characterization

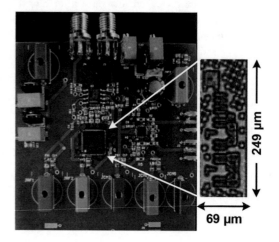

the output of active RC and proposed filter. The resistors used in both the designs
are 5 kΩ. Simulation shows around 10 dB better noise performance of the proposed
filter.

5.3 Measurements

The proposed third order active inverter-based filter is fabricated in a IBM 65 nm
CMOS process. The area occupied by the filter is 0.0175 mm^2, of which 94.5 %
are filter capacitors. An open-drain source-degenerated common source amplifier is
used to drive the output off-chip. The die was mounted in a QFN package on a PCB
and tested with high-performance discrete differential opamps for input and output
buffering and 50 Ω interfaces (Fig. 5.19). Phase and amplitude-matched 50 Ω baluns
were used to interface with single-ended equipment. The measurement results are
discussed below.

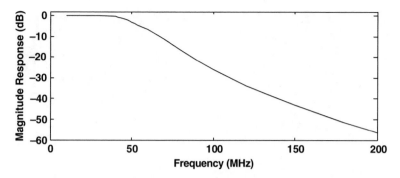

Fig. 5.20 Measured frequency response of the third-order Butterworth filter

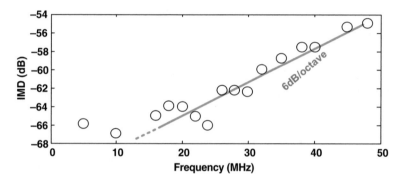

Fig. 5.21 Measured IMD for $400\,\mathrm{mV}_{pp\text{-}diff}$ with 0.5 MHz tones separation

A Butterworth response and cutoff frequency of 58.7 MHz are measured in the magnitude response plot shown in Fig. 5.20. Global variation of R and C caused equal deviation from the 50 MHz target for all five chips measured. The filter core and common-mode circuits consume 1.3 mA at 1.2 V. Additionally, the measured integrated output noise is $648\mu\,V_{rms}$. This can be reduced by using smaller resistors and larger capacitors. Figure 5.21 show the intermodulation distortion plot across the pass band frequency of the filter. Figure 5.22 shows a screen capture of a particular test result for two-tone inputs at 39.5 and 40 MHz. The IM_3 products are 57.6 dB down as indicated in the plot. As can be seen from the trend line in the figure, the IMD increases with a first-order slope (6 dB/oct.) due to amplifier BW limitations.

Table 5.1 summarizes our design and compares it with other published state-of-the-art filters with cutoff frequencies in this range. For a fair comparison, we have divided the area by the order of the filter. As area is dominated by the passives, it will be fairly independent of technology. From Table 5.1, we see that this work occupies the smallest area by far. This is primarily due to reuse of the filter capacitors as compensation capacitance. The design has a fairly high IIP_3 and SFDR, even near f_c,

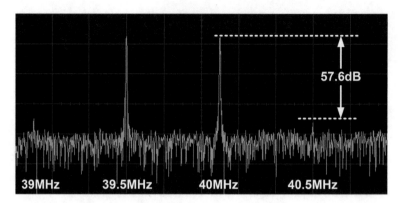

Fig. 5.22 Measured IMD with tones at 40 MHz separated by 0.5 MHz

Table 5.1 Summary and comparison of proposed filter with prior art

	This work	[31]	[28]	[32]	[25]	[33]
Tech (nm)	65	180	180	180	90	130
Supply (V)	1.2	1.8	1.8	1.8	0.9	1.5
Power (mW)	1.56	4.1	4.5	0.47	19.1	11.25
Area/N (mm^2)	0.0057	0.065	0.04	0.035	0.073	0.04
Order (N)	3	4	5	6	4	5
BW (MHz)	58.7	10	20	20.3	30	19.7
Noise (nV/$\sqrt{\text{Hz}}$)	80.7	7.5	225	65.4	26.0	30.0
IIP$_3$ near f_c (dBm)	33	12	41.5	13.6	34.3[a]	18.3
SFDR[a] near f_c (dB)	55.8	60.9	60.3	47.4	65.5[b]	55.1
FOM (aJ)	23.1	82.9	42.0	70.2	44.9	353

[a]SFDR (dB) $= \frac{2}{3}\left[\,\text{IIP}_3\,(\text{dBm}) - P_{\text{noise}}\,(\text{dBm})\,\right]$
[b]Estimating SFDR from DR $= 65.5\,\text{dB}$ and low frequency THD $= 65.2\,\text{dB}$

however, these numbers are even more impressive when we note that our bandwidth is 2X or better than others. As shown in Fig. 5.21, the nonlinearly degrades with increasing frequency, as expected, but our relaxed amplifier UGF does not cause much detriment.

For overall comparison, the figure-of-merit below is used [28].

$$\text{FOM} = \frac{\text{Power}}{\text{N} \cdot \text{SFDR} \cdot \text{BW}}$$

5.4 Conclusion

We present a low pass filter based filter architecture rather than the integrator based architecture in current mode. This reduces the number of low impedance nodes and thus giving power and noise advantage of corresponding active RC filters. Further we present a circuit technique that combines the advantages of active-RC and G_m-C filters. Driving the load in open loop to increase speed while using an independent feedback loop to improve linearity ($IIP_3 = +33$ dBm) enables an FOM=23.1 aJ. This makes the design attractive for baseband filtering in portable wireless transceivers. This circuit occupies very little area (6X lower than the state-of-art) compared to the existing active-RC filters, thanks to the self compensation of amplifiers. This design also provides an additional degree-of-freedom for filter Q control that is not impacted by finite OTA gain or G_m-C output impedance like traditional designs.

Chapter 6
All MOSCAP Based Continuously Tunable Filter

The prevalence of multiple wireless standards and software-defined radios [34–36], requires baseband channel select filters with a wide tuning range. Active filters are often implemented using Gm-C, active-RC, or Gm-OTA-C techniques [29]. Gm-C filters can be tuned continuously and are capable of high-speed operation due to their open-loop nature. Open-loop operation also means that the linearity of Gm-C filters is limited by the transistor's inherent non-linearity. Increasing the overdrive voltage improves the linearity at the expense of higher power, higher noise and reduced signal swing. Techniques that attempt to linearize the Gm results in higher noise. An alternate method to improve Gm-C filter linearity is through the use of source follower based Gm-C filters [32]. The major drawback, however, is the limited output swing and this poses a serious limitation on Gm-C filters at lower technologies.

Active-RC filters use OTA-based integrators, resulting in a linearity that is only limited by the swing at the virtual ground node of the amplifier [28]. However, designing high-frequency opamp RC filters is problematic, since the gain-bandwidth product of the opamps in an active RC biquad must be much larger than twice the product of the highest quality pole pair of the filter transfer function [37]. These problems become even more significant at lower technology nodes where the poor output impedance and lower supply voltage affects the quality factor of the integrator. Further, continuous tuning of active-RC filters is difficult. Banks of resistors and/or capacitors [33, 38–43] can be employed in active-RC filters to offer programmability; however, they occupy relatively large silicon area. In addition, the transistors switches (quasi-static switches) within the banks are associated with finite nonlinear resistances. The opamps in active-RC has to be compensated at maximum frequency resulting in over compensation at lower frequency. In a Active-Gm-RC approach, the biquadratic cell presents a closed-loop structure that exploits the opamp frequency response in the filter transfer function. This corresponds to operating with an opamp unity-gain-frequency comparable with the filter pole

© Springer International Publishing AG 2017
R.K. Palani, R. Harjani, *Inverter-Based Circuit Design Techniques for Low Supply Voltages*, Analog Circuits and Signal Processing,
DOI 10.1007/978-3-319-46628-6_6

frequency and hence minimizes the power consumption with respect to other closed loop structures. The input and output impedances of the Gm-C filter are high and it requires a buffer to drive the next stage in the chain if the load cannot be absorbed into the filter capacitance. Although active-RC filters has low output impedance, they require the previous stage to drive their resistive input.

In this chapter we propose an inverter based filter architecture [44] which uses only the non-linear MOS capacitor (MOSCAP) as the filter capacitors [17]. The filter is tuned continuously by varying the capacitance of the MOSCAP. Linearity is achieved by reducing the swing across the MOSCAPs. A low pass filter based filter design is proposed rather than the traditional integrator based design for lower noise and ease of design in lower technologies. The superior transconductor efficiency of the inverter translates to lower power and lower noise of the filter. Semi constant current biasing of the inverter is used to reduce PVT variations. The proposed third order channel select filter fabricated in TSMC's 65 nm general purpose technology achieves an IIP_3 of $+22$ dBm while drawing 4.2 mA from a 1.1 V supply and occupies an area of 0.01 mm^2. It is tunable from 34 to 314 MHz without any capacitor bank achieving 9X tuning range.

6.1 Filter Architecture

This filter takes the input at the gate of an inverter and delivers the output at the output impedance of a negative feedback circuit. This features offers negligible loading on the front end circuits thereby easing system integration. This filter is also based on low pass filter based architecture. The design implemented is fully differential and single ended schematic is shown in Fig. 6.1. The OTA (Fig. 6.2) used is realized as a cascade of scaled gm-cells and each gm-cell (Fig. 6.3) is realized as a semi constant current biased inverter. A third order filter is obtained by cascading first order and biquad filters. The negative feedback creates the complex poles of the biquad [17] The biquad is realized as a cascade of low pass filters and a gain element in negative feedback as shown in Fig. 6.1.

6.1.1 Root Locus

The root locus of the poles with gain is shown in Fig. 6.4. Initially (when gain K = 0) the system has two real poles located at P_2 and P_3. When the gain increases, poles move toward each other until they meet on the real axis and then they split and move in the complex plane as shown in Fig. 6.4. The gain in the negative feedback is decided by the location of poles on complex plane as dictated by the filter. The low pass filters are obtained by cascading two first order systems and the ratio of the gm cells (G3, G4 and G5, G6) provides the gain. The resistive divider R1 along with the ratio of gm cells allows for non-integer gain. Each first order system consists of a resistor (R and 2R for biquad) and capacitor C.

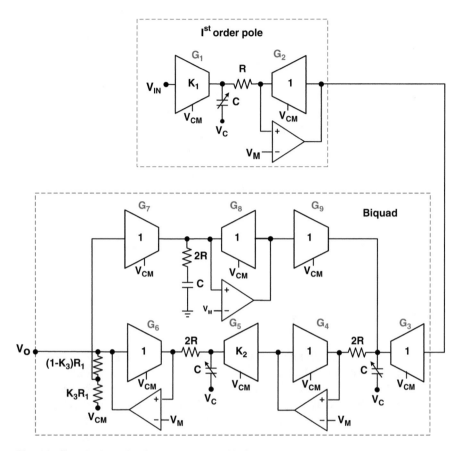

Fig. 6.1 Circuit schematic of the proposed tunable filter

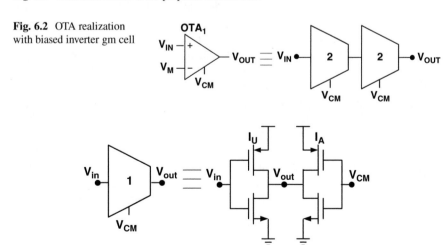

Fig. 6.2 OTA realization with biased inverter gm cell

Fig. 6.3 Circuit schematic of the gmcell

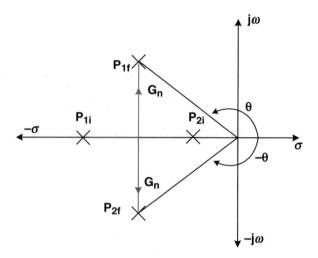

Fig. 6.4 Root locus of the two poles system

6.1.2 First-Order System

Figure 6.5 shows the circuit's schematic of the first order system obtained by steering current across RC circuit. The input is converted to current by the input-conductance (G_1) of K_1 size and is then low pass filtered by the RC of the first order system. The RHS of resistor R is a virtual ground. The low passed current I_R through R, which is proportional to the capacitor voltage, is absorbed by another identical output trans-conductance (G_2) of unit size using an OTA in negative feedback. The low value of R and higher value of capacitor reduces the swing across the capacitor and also helps in compensation as explained later. This facilitates the use of nonlinear MOSCAPs and enables continuous tuning of the capacitor. Additionally, the reduced swing at G1's output reduces the output impedance (g_{ds}) mismatch between the two gm cells without introducing any noise penalty owing to the current input node. The input, output and swing across the MOSCAP is shown in Fig. 6.6 for a 10 MHz input. A 400 Ω resistor and a 8 pF capacitor is used in this design. The output swing (red) is twice that of the input swing (blue) due to the ratio of gm cells G_1 and G_2 (2 in this design). The swing across the MOSCAP is 102 mV in this design due to the choice of resistor and capacitor selected here. The 6 dB reduction of swing from input on the MOSCAP leads to 18 dB improvement in third order distortion due to its nonlinear capacitance. Unlike conventional active RC filters, a lower filter resistance does not reduce OTA-G2's gain because it is in series with the capacitor C. In order to study the non-linearity cancellation between the gm cells (G_1 and G_2), two situations are discussed.

Fig. 6.5 Circuit schematic of the first order system

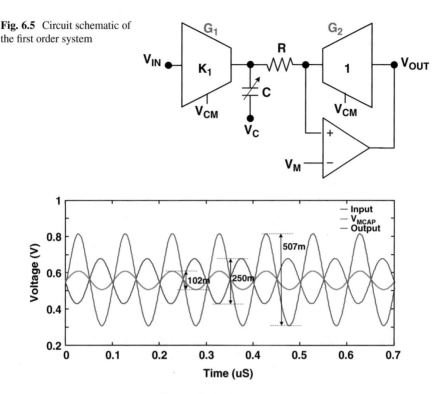

Fig. 6.6 Voltage swing across various nodes in the first order system

6.1.2.1 Low Frequency Input

Assume the signal frequency is low and all the harmonics are within the cut off frequency of the *RC* filter. The gm cell's current is a non linear function of the input voltage and is represented as

$$G_m(x) = a_0 + a_1 x + a_2 x^2 + a_3 x^3 \tag{6.1}$$

The sizes of the transconductances G_1, G_2 are K_1 and one times the size of the unit gm cell. Hence from Eq. (6.1), we have $G_1 = K_1 G_m$ and $G_2 = G_m$. The impedance of the capacitor is large at low frequency and hence almost all the current is absorbed by the gm cell G_2 giving the output voltage V_{OUT}.

$$G_1(V_{IN}) = G_2(V_{OUT}) \tag{6.2}$$

$$K_1 a_0 + K_1 a_1 V_{IN} + K_1 a_2 V_{IN}^2 + K_1 a_3 V_{IN}^3 = a_0 + a_1 V_{OUT} + a_2 V_{OUT}^2 + a_3 V_{OUT}^3 \tag{6.3}$$

Equating the currents of the gm cells G_1 and G_2 as shown in Eq. (6.3) we obtain the output voltage as

$$V_{OUT} = \frac{a_0}{a_1}(K_1 - 1) + K_1 V_{IN} + \frac{a_2}{a_1}(K_1 V_{IN}^2 - V_{OUT}^2) + \frac{a_3}{a_1}(K_1 V_{IN}^3 - V_{OUT}^3) \quad (6.4)$$

In a differential implementation the even harmonics cancel giving the output voltage as

$$V_{OUT,DIFF} = K_1 V_{IN,DIFF} + \frac{a_3}{a_1}(K_1 V_{IN,DIFF}^3 - V_{OUT,DIFF}^3) \quad (6.5)$$

The third order coefficient (a_3) is minimized by the scheme of inverter biasing and hence the output voltage is approximated as $K_1 V_{IN,DIFF}$. Substituting this voltage in Eq. (6.6), we obtain the output voltage as

$$V_{OUT,DIFF} = K_1 V_{IN,DIFF} + \frac{a_3}{a_1}(K_1 - K_1^3)V_{IN,DIFF}^3 \quad (6.6)$$

If the conversion gain (K_1) is one, $V_{OUT,DIFF} = K_1 V_{IN,DIFF}$. Since the conversion between voltage and current is done by identically sized gm cells (G1 and G2) any non-linearity of these gm cells is canceled completely. However with gain in the gm cells, the cancellation is partial and the residue IIP_3 is given by

$$IIP_3 = \sqrt{\left|\frac{4}{3(1 - K_1^2)}\frac{a_1}{a_3}\right|} \quad (6.7)$$

6.1.2.2 High Frequency Input

Input signal frequency is high enough such that higher order harmonics fall out of band of the RC circuit. The input to the first order system is assumed as sum of closed spaced sine waves and is given by

$$V_{IN} = A_1 sin(\omega_1 t) + A_2 sin(\omega_2 t) \quad (6.8)$$

Since the higher frequencies are filtered by the RC circuit, the output will have these components as Eq. (6.9)

$$V_{OUT,DIFF} \approx K_1 V_{IN,DIFF} + (K_1 - K_1^3)V_{IN,DIFF}^3 + \frac{1}{4}(A_1 sin(3\omega_1 t) + A_2 sin(3\omega_2 t))$$

$$+ \frac{3}{2}(A_1 sin((\omega_1 + 2\omega_2)t) + A_2 sin((\omega_2 + 2\omega_1)t)) \quad (6.9)$$

Since the high frequency signals get filtered by subsequent filtering stage, the IIP_3 of the first order system remains unchanged as Eq. (6.7).

6.1.2.3 Compensation

The output current i_{out} is absorbed by the gm cell G_2 using negative feedback with operational trans conductance amplifier (OTA_1). OTAs are realized as a cascade of two double-size gm cells (G_{o1} and G_{o2}). Each gm cell is realized as biased inverter with the meta- stable voltage V_M. The negative feedback loop consists of three stages of biased inverters. This negative feedback has to be compensated over the entire tuning range. The gm cell G_2 pumps current into the series resistor and capacitor combination creating a left half plane zero. If the input resistor is smaller than the output impedance of inverter and the filter capacitor is greater than the parasitics, the amplifier can be compensated with this zero. The loop is broken at the OTA_1 input to study the loop dynamics as shown in Fig. 6.7. A series combination of resistor(R_{c1}) and capacitor(C_{C1}) is connected at the output of gm cell (G_{o1}) in OTA_1 to aid in compensation. In Fig. 6.8 r_{o1}, r_{o2} are the finite output impedances of transistors and C_{p1}, C_{p2} and C_{p3} are the parasitic capacitance at the gm-cells output. The loop gain can be derived as Eq. (6.10)

$$\frac{V_{TO}}{V_T} = A_{DC}\frac{V_1}{V_T}\frac{V_2}{V_1}\frac{V_{TO}}{V_2} \quad \text{where} \tag{6.10}$$

$$A_{DC} = 4gm^3 r_{o1}^2 r_{o3} \tag{6.11}$$

$$\frac{V_1}{V_T} = \left(\frac{(1 + sR_{c1}C_{c1})}{\frac{1}{r_{o1}} + s(C_{p1} + C_{c1} + \frac{R_{c1}}{r_{o1}}C_{c1}) + s^2 R_{c1}C_{c1}C_{p1}}\right) \tag{6.12}$$

$$\frac{V_2}{V_1} = \left(\frac{r_{o1}}{(1 + sC_{p2}r_{o1})}\right) \tag{6.13}$$

$$\frac{V_{TO}}{V_2} = \left(\frac{(1 + sR_{c1}C_{c1})}{\frac{1}{r_{o3}} + s(C_{p1} + C_{c1} + \frac{R_{c1}}{r_{o1}}C_{c1}) + s^2 R_{c1}C_{c1}C_{p1}}\right) \tag{6.14}$$

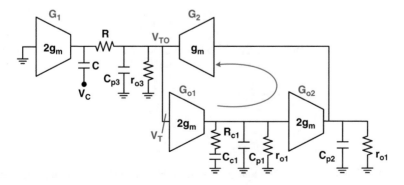

Fig. 6.7 First order system with parasitics

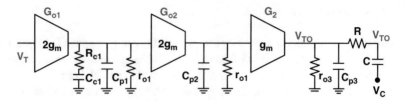

Fig. 6.8 Circuit schematic to evaluate loop gain

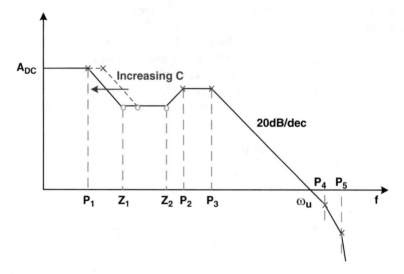

Fig. 6.9 Bode plot of the loop gain

The system has five poles and two zeros. The bode plot of these poles and zeros are given in Fig. 6.9. The poles, zeros and unity gain frequency (ω_u) can be approximated as

$$Z_1 = -\frac{1}{RC} \qquad Z_2 = -\frac{1}{R_{c1}C_{c1}} \tag{6.15}$$

$$P_1 \approx -\frac{1}{r_{o3}C} \qquad P_2 \approx -\frac{1}{r_{o1}(C_{p1}+C_{c1})} \tag{6.16}$$

$$P_3 \approx -\frac{1}{r_{o1}C_{p2}} \qquad P_4 \approx -\frac{1}{R_{c1}C_{p1}} \tag{6.17}$$

$$P_5 \approx -\frac{1}{RC_{p3}} \qquad \omega_u \approx \frac{A_{DC}}{r_{o1}C_{p2}} \tag{6.18}$$

Since the resistors R_{c1}, R and capacitors C_{c1}, C are chosen to be smaller and larger by design, poles P_4 and P_5 are at high frequency and their impact on the stability is negligible. There are three poles and two zeros within the unity gain frequency. By appropriately positioning zeros, the loop can be compensated. It is to be noted that the filter capacitor creates both a pole and a zero and hence with tuning both move in the same direction without affecting the stability. Further lower value of filter resistor (R) and higher capacitor (C) does not increase the power of the loop unlike the traditional active RC filters. It is seen from Eq. (6.18) that the unity gain frequency of the loop in a well compensated system is independent of the large filter capacitor enabling high frequency operation of the loop.

6.1.3 Third Order Filter

The biquad is realized as a cascade of low pass filters and a gain element in the negative feedback as shown in Fig. 6.1. The negative feedback creates the complex poles of the biquad [17]. The low pass filters are obtained by cascading two first order systems and the ratio of the gm cells (G3, G4 and G5, G6) provides the gain. The resistive divider R1 along with the ratio of gm cells allows for non-integer gain. Each first order system consists of a resistor (R and 2R for biquad) and capacitor C. Third order filter is obtained by cascading the first order system and the biquad. The transfer function of filter is given by

$$\frac{V_O}{V_{IN}} = \frac{K_1 K_2 K_3}{1 + K_2 K_3} \frac{\omega_o}{(s + \omega_o)} \frac{\omega_o^2}{(s^2 + \frac{w_o}{Q}s + \omega_o^2)} \tag{6.19}$$

$$\text{where} \quad Q = \frac{\sqrt{1 + K_2 K_3}}{2}, \qquad \omega_o = \frac{1}{RC} \tag{6.20}$$

6.2 Biasing and CMFB

The inverters used in this design are biased using semi constant current biasing technique as described in Biasing chapter. The NMOS in the unit inverter is biased at $40\,\mu\text{A}$ and the PMOS current is set by the biasing network. The common mode feedback circuitry is shown in Fig. 6.10. The voltage nodes V_P and V_M are the differential modes in the circuit. The resistors (10k) in Fig. 6.10 senses the common mode voltage and adjusts the common mode to the meta stable voltage V_M of the semi constant current biased inverter using OTA in negative feedback. The OTA in this design is realized using a five transistor differential pair. We used this CMFB at two places in filter, one after the first order pole and other at the output of biquad.

Fig. 6.10 Common mode
feedback circuit for tunable
filter

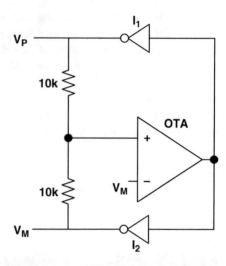

Fig. 6.11 Micrograph of
proposed tunable channel
select filter

6.3   Measurement Results

The proposed filter is fabricated in TSMC 65 nm general purpose technology and
its micrograph is shown in Fig. 6.11. This design is modular since all the gm cells
are multiples of a unit biased gm cell. The area occupied by the filter is $0.007 \, mm^2$,
of which 94.5 % are filter capacitors. A source follower is used to drive the output
off-chip. The source follower's gain is calibrated on chip and is used in measuring
filter gain. The die is mounted in a QFN package on a PCB and tested with 50 Ω
interfaces. Phase and amplitude-matched 50 Ω baluns were used to interface with
single-ended equipment. The measurement results are discussed below.

Figure 6.12 shows the measured magnitude response of the channel select filter.
The filter has a DC gain of 3 dB and is tunable from 34 to 314 MHz (9.2x) with the
aid of the control voltage (Vc). A flat response is observed over the pass band with
peaking of 0.3 dB only at the highest frequency. Figure 6.13 shows the measured
OIP3 of the filter with tones at 260 MHz with 1 MHz offset at a control voltage

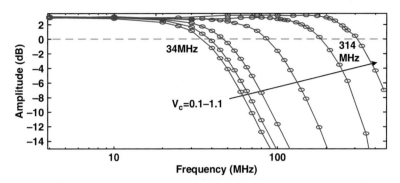

Fig. 6.12 Measured magnitude response of proposed filter

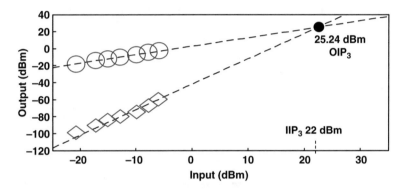

Fig. 6.13 Measured IIP3 of proposed filter at 260 MHz input

1.1 V (314 MHz). Figure 6.14 shows the snap shot of a IMD measurement with 260 MHz input and at control voltage of 1.1 V. The IMD depends on the MOSCAP nonlinearity, gm-cell non-linearity cancellation and also on the g_{ds} non-linearity. All the gm-cells (biased-inverters) are linearized using biasing technique and they can support a larger swing. Further swing across the non-linear MOSCAPS are reduced to facilitate the capacitance linearity. Hence we obtain an OIP_3 of +25.24 dBm and IIP_3 of +22 dBm in this design. Additionally, the measured integrated output noise is 620 μVrms over the bandwidth of 314 MHz when the control voltage is 1.1 V. The power of the filter is fairly independent of the tuning. It draws 4.2 mA from a 1.1 V supply over the entire tuning range. The power includes the filter core power and the biasing power. This is because frequency tuning does not change the power consumption of the gm cells or the gain and unity gain frequency of the OTA. The filter output is followed by an on-chip source follower driving an off chip 50 Ω load whose attenuation, measured to be 31 dB, is calibrated from the measurements. Figure 6.15 shows the IMD and OIP3 measured at the band edge over the tuning range. The bottom of the figure shows the intermodulation test done for two tone inputs at 260 and 261 MHz. IMD is measured by placing two −8 dBm

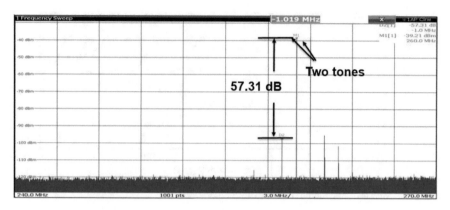

Fig. 6.14 Measured IM_3 at 260 MHz, $V_c = 1.1$ V, with tones separated by 1 MHz

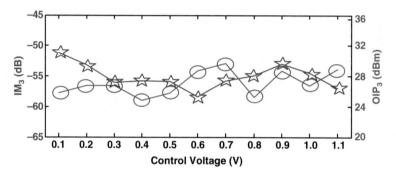

Fig. 6.15 Measure IM3 and OIP3 at band edge frequency

tones at band edge with 1 MHz offset. The intermodulation distortion varies from 51.2 to 57 dB while OIP3 varies from 21 to 25.24 dBm over the entire tuning range. Further in order to verify the sensitivity of the circuit towards temperature and the power supply, IMD is measured across industry standard temperature ranging from −40 to 80 °C and power supply variation of ±100 mV. Figures 6.16 and 6.17 show IMD measured over temperature and power supply when the control voltage is set to Vc = 1 V. IMD at the band edge is measured across 15 chips and varies from −52 to −59 dB for Vc = 1.1 V. IMD varies from −51.5 to −58 dB for over a 200 mV variation in supply voltage. IMD varies only 5 dB across temperature. IMD is measured across 15 chips and is given in Fig. 6.18. This is due to the modular design, biasing techniques and tolerance of the circuit to component mismatch. Table 6.1 shows the performance summary and compares it with other published state-of-the-art filters with cutoff frequencies in this range. Although the filter uses large filter capacitors of 8 pF, it occupies the smallest area of 0.007 mm^2 due to the use of high density MOSCAPs. For a fair comparison, we have divided the area by the order of the filter. As area is dominated by the passives, it will be fairly independent of technology. From Table 6.1, we see that this work occupies

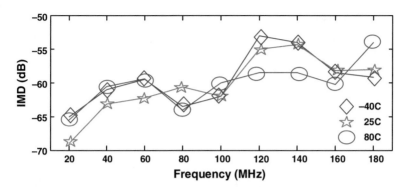

Fig. 6.16   Measure IM3 with frequency at $Vc = 1$ V over temperature

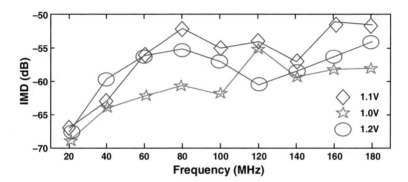

Fig. 6.17   Measure IM3 with frequency at $Vc = 1$ V with power supply

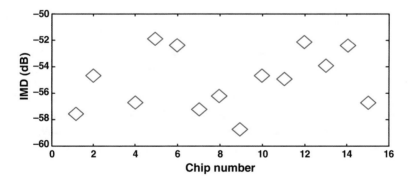

Fig. 6.18   Measure IM3 at band edge across 15 chips

Table 6.1 Performance summary and comparison with prior art

	[45]	[46]	[25]	[47]	[48]	This work
Tech (nm)	180	65	90	90	180	65
Supply (V)	1.2	0.6	0.55–0.9	1	1.8	1.1
Power (mW)	4.7–11	26.2	1.9–1.9	4.4	90	4.6
Area/N (mm²)	0.076	0.095	0.073	0.039	0.035	0.002
N	3	4	4	8	5	3
BW (MHz)	0.5–20	70	7–30	8.1–13.5	200–500	34–314
Tuning Ratio	40	1	6.58	1.67	2.5	9.23
IRN (nV/$\sqrt{Hz}$	425–12	44	33	75	18	25
IIP_3 (dBm)	19.0	32.8 [a]	29.6 [a]	22.1	26.5	22.4
FOM (aJ)	151	117	126	238	135	61
FOMT	183	182	177	173	164	186

[a] THD is used to estimate IIP_3 at band edge

the smallest area by far. This is primarily due to realization of filter capacitors as MOSCAPs and its reuse as compensation capacitance. The design has a fairly high IIP3 and SFDR, even near fc, however, these numbers are even more impressive when we note that our bandwidth is 2X or better than others. The filter has the largest tuning range among active-RC filters and is tolerant to PVT variations. The filter is compared with state-of-art using the figure of merit FOM [17] and FOMT [45] including the tuning range.

$$\text{FOM} = \frac{\text{Power}}{\text{Order x SFDR x BW}} \tag{6.21}$$

$$\text{FOMT} = 10\log10\left(\frac{\text{Order x SFDR x } f_o \text{ x tuning}}{\text{Power}}\right) \tag{6.22}$$

where SFDR is defined as the signal to noise ratio when the power of the third-order inter-modulation distortion term equals to the noise power, f_o is the geometrical mean of the cutoff frequency in the unit of Hz, tuning is the tuning ratio of the low pass filter, power is the geometrical mean of power in the unit of Watt and BW is the cut off frequency of the filter in Hz. This work achieves the highest FOMT and lowest FOM while supporting higher operating frequencies and occupying 17.5x smaller area compared to prior art.

6.4 Conclusion

This chapter presents an efficient solution to design flexible analog filter circuits. This is demonstrated with a CMOS 65-nm implementation of a tunable low-pass filter using only MOSCAPs. Although filter uses MOSCAPs, it achieves IIP_3 of

22 dBm at the highest tuning frequency. Further all the negative feedback circuits are self compensated using the filter resistor and capacitor resulting in low power of 4.6 mW and 17.5x smaller area. Due to the biasing of inverter with semi constant current biasing and owing to modular design, IMD varies only by 6.5 dB over 200 mV variation in power supply and 5 dB across temperature. The filter achieves the highest figure of merit among the state of art published filters.

Chapter 7
ADC

Over the last two decades wireless mobile handsets have become increasingly more complex and smart-phones are largely de rigueur now. Mixed signal SoC applications require the integration of the analog front end with digital signal processing and control blocks. Low power and small area ADCs with 8–10b accuracy operating at several tens of MS/s are critical blocks within such systems. Most ADC applications can be classified into four broad market segments [49]: (a) data acquisition, (b) precision industrial measurement, (c) voiceband and audio, and (d) high speed (sampling rates greater than 5 MS/s). A very large percentage of these applications can be filled by successive-approximation (SAR), sigma-delta ($\Sigma - \Delta$), and pipelined ADCs. Applications determine the required resolution and hence the architecture (http://www.ti.com/europe/downloads/choose/). Audio applications, temperature and pressure sensors requires higher resolution but they have low bandwidth requirements. On the other hand applications like communications, Defense, Imaging and testing requires resolution from 9 to 16 bit with bandwidth from 1 M to 1 GHz as shown in Fig. 7.1.

The application determines the ADC architecture as shown in Fig. 7.2. Further time interleaving is done to extend the bandwidth of the ADCs. This work is aimed at building a high speed ADC for communication applications with sampling rate greater than 100 MS/s. This leaves us with the architecture choice of pipeline and time-interleaved SAR ADC. A pipelining data converter uses a cascade of individual stages which each resolving bits. The simplest pipeling ADC is a two stage design with the first stage resolving MSB and the second stage resolving LSB. The residue amplifier is used to amplify the residue of the first stage to be used by second stage. The use of residue amplifier also allows to sample and hold the residue, or in other words pipeline the stages of the ADC. This allows us to use at higher sampling rate. Although amplifiers ease the design of subsequent stages of pipelined ADC after it, there design in lower technologies is becoming difficult due to lower power supply and output impedances.

© Springer International Publishing AG 2017   103
R.K. Palani, R. Harjani, *Inverter-Based Circuit Design Techniques for Low
Supply Voltages*, Analog Circuits and Signal Processing,
DOI 10.1007/978-3-319-46628-6_7

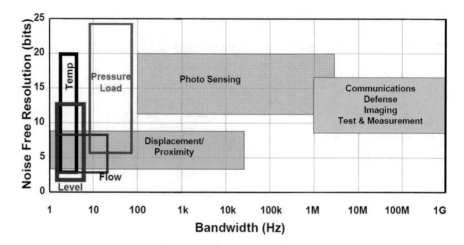

Fig. 7.1 Real world versus bandwidth

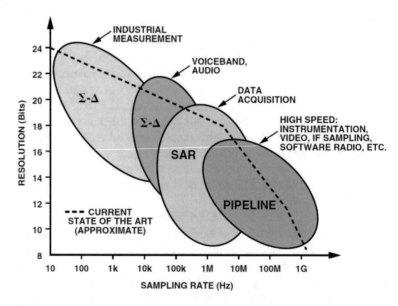

Fig. 7.2 Architectural choice based on applications

The SAR ADC has become increasingly popular for medium resolution applications as it is based on device switching rather than amplifying. Further the SAR
ADC requires less hardware and scales well with technology. Further any offset
and gain error in comparator becomes the offset and gain error to the entire ADC.
SAR ADC typically suffer from two problems, higher input capacitance and high
frequency clock. One fundamental bottleneck of SAR ADCs is metastability due to
finite comparator settling time as sample rates increase. This requires higher gain,
which in turn translates to higher power consumption.

However, even recent SAR ADCs that have good energy efficiency in scaled technologies have had relatively large input capacitances, on an average of 1–1.8 pF [18, 50], because in traditional SAR topologies the input signal is sampled onto the DAC capacitor array itself. The DAC array could potentially be made smaller by using custom capacitors [51, 52] or by using split array DACs [53, 54]. However, these are likely to require DAC calibration or carefully designed using a dedicated CAD tools for process mismatch characterization.

SAR ADCs can be of synchronous and asynchronous design. In synchronous design each bit is resolved using a global high frequency clock which is determined by the time required to charge the MSB capacitor. This gives a over design for resolving a small LSB capacitors. Asynchronous SAR ADCs were introduced to try to solve this problem by allocating more time for resolving MSBs but results in more complex clocks and added synchronization [55]. Further the sampling time is determined by the nonlinearity of the sampler. Lowering the sampling capacitor or increasing the switch size increases the sampler bandwidth which reduces its nonlinearity. The subthreshold leakage, clock feed through and charge injection increases with the switch size. The lower sampling capacitor lowers the driver power and also reduces the sampler non linearity.

Synchronous SAR ADCs suffer from high frequency clock requirements. In synchronous SAR ADCs each bit gets resolved in one high frequency clock period and sampling time is typically half the sample clock period. Hence the minimum clock frequency (f_{min}) required is given by $f_{min} = 2Nf_{nq}$, where N is the number of bits and f_{nq} is the Nyquist frequency. For example if the sampling time is one clock period and if we time interleave synchronous SAR ADCs by N + 1, the clock frequency required by this ADC is the Nyquist frequency. The sampling time reduction to the time required to resolve 1 bit requires the design of a high frequency sampler. The design of high frequency samplers is simplified significantly by reducing the size of the sampling capacitor.

While the core ADC power has scaled with technology, the power of the input buffers and reference generators, that have to drive these ADCs, have continued to increase due to the limited voltage headroom and lower impedances used. The power consumed by the drivers/amplifiers is normally higher than the power consumed by the ADC itself at these scaled technologies [17, 18, 56]. This is primarily due to the fact that the drivers rely on linear gain that is obtained by utilizing negative feedback while ADCs rely on nonlinear comparator gain based on positive feedback. As the driver power scales with the load capacitance, i.e., the input capacitance of the ADC, some earlier efforts have been taken to reduce the input capacitance of the ADC either by using architectures like pipelined SARs or by using split array capacitance but they all require additional calibration steps [26, 57–59]. Single ended SAR designs compare the input to the DAC reference voltage. However in fully differential designs, the input difference has to be subtracted from the DAC reference difference [43]. In the proposed design, the DAC feedback signal is subtracted from the input signal within the preamplifier in the current domain.

In this chapter, we present a time-interleaved synchronous SAR ADC [60] that has low input capacitance of 133 fF and can accommodate up to 10X interleaving (max 1.1 GS/s) without increasing the input capacitor. This is achieved by using a separate sampling capacitor instead of using the DAC array for sampling and by performing the subtraction of the input and DAC voltages in the current domain within the preamp rather than by using traditional charge based subtraction in the DAC array [61]. We are able to use a smaller capacitor for sampling than is required for the DAC array because the DAC array capacitor size is set by device matching, which is more restrictive than required for sampling, which is set by the kT/C limit, by over two orders of magnitude [62].

The two problems with SAR design, high frequency clock requirement more than sampling frequency and higher input capacitance is solved by time interleaving and using a separate sampling capacitor rather than the DAC array in this work. Further in the fully interleaved 10X version, the ADC driver sees at constant capacitive load of 133 fF.

7.1 ADC Architecture

The proposed 2X time-interleaved synchronous SAR ADC architecture is shown in Fig. 7.3. It is setup for full 10X interleaving (N + 1) but only two of these channels were implemented in this prototype. A ring oscillator based counter is used to generate 10 clock phases of 909 ps pulse-width from a 1.1 GHz clock. The fifth and tenth phase are used for sampling for the sub-ADCs as shown in Fig. 7.3 and remaining nine phases are used to resolve the bits. Due to the small sampling capacitor only one phase is devoted for sampling and additional nine phases are used to resolve the 9 bits resulting in a doubling of the speed for the 2X interleaving.

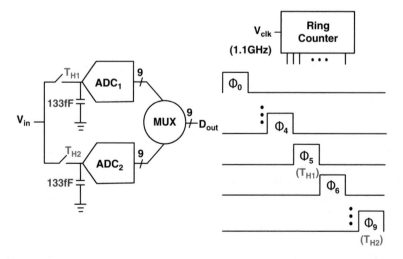

Fig. 7.3 Overall architecture and timing diagram for the time-interleaved ADC

The input signal is sampled onto a 133 fF capacitor using a constant-V_{gs} sampler. The preamp subtracts the differential reference feedback voltage from the differential sampled voltage. It also functions to reduce the comparator kick back and offset. The strong arm comparator resolves this difference to logic levels and is fed to the SAR logic which generates the corresponding DAC voltages for subsequent steps of the SAR process.

7.2 DAC Design

The single ended circuit schematic for SAR sub-ADC is shown in Fig. 7.4. The actual design implemented is fully differential. The DAC utilizes a binary-weighed capacitor bank except for LSB-1 and LSB as a compromise between area and linearity. In this process the top-plate parasitics are negligible and the bottom-plate parasitics are ≈4 %. The LSB capacitor (1.25 fF) is realized as a series combination

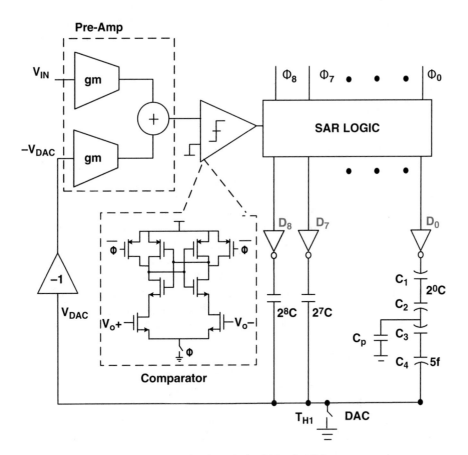

Fig. 7.4 Simplified single ended circuit schematic for SAR sub-ADC

of capacitors C_{1-4} as shown in Fig. 7.4. The top plates of C_1 and C_2, C_3 and C_4 are connected together. This reduces the parasitics at the inter-connection of C_1, C_2 and C_3, C_4. However, the bottom plates of C_2 and C_3 create the parasitic capacitance C_p which results in less than 0.5LSB error in simulations. The LSB-1 capacitor is realized as a series combination with the top-plates at the center, i.e., little parasitic impact. This split array technique reduced the DAC array capacitance from 2.55 to 0.64 pF ($\approx$ 4X reduction). The parasitic capacitance of the overall DAC array results in benign gain and offset errors.

7.3 Sampler Design

A constant-V_{gs} sampler is used to sample the input. Increasing the size of the sampling switch reduces the switch non-linearity but increases the subthreshold leakage and nonlinear parasitic capacitance particularly when the sampling capacitor is small as in our case. Subthreshold leakage is partially canceled by using a cross-coupled dummy sampling transistor with a grounded gate. The impact of the non-linear capacitance is minimized by appropriately sizing the preamp input transistor as described in Sect. 7.5.

7.4 Preamp Design

Figure 7.5 shows the fully differential preamplifier circuit. Inverters are used as the g_m cells for converting voltage to current of the sampled input voltage and the DAC voltage. The resistors are used to maintain the output common mode of the preamp to approximately mid-V_{dd}. Subtraction of the input and the DAC voltage is realized by swapping the differential wires and adding the outputs of the g_m cells.

Assume the input differential voltage is −0.8 V (V_{IP} = 100 mV, V_{IM} = 900 mV). Signal input is designated as V_{IP} and V_{IM} and DAC input as V_{DP} and V_{DM} in Fig 7.5. The inverters I_1–I_4 (gm cells) convert the voltages to current and sum them in the current domain. When V_{IP} is 100 mV and V_{IM} is 900 mV, in the first step of conversion both the DAC voltages are at 0.5 V, resulting in the triode operation of the PMOS and NMOS in inverters I_1 and I_2. So the outputs V_{OM} and V_{OP} will be close to power supply and ground generating larger difference to be detected by the comparator.

In the second step the DAC voltages V_{DM} is 0.75 V and this makes the NMOS of I_2 stronger thereby pulling the V_{OM} towards ground. Similarly the DAC voltage V_{DP} is 0.25 V and makes the PMOS of I_4 stronger thereby pulling V_{OP} towards the power supply. With few steps (two to three conversion steps) the output voltages are near 0.5 V ensuring that all the transistors enter the saturation region. A step by step process is explained in the example next for clarification. Assume the input V_{IP} is 100 mV and V_{IM} is 900 mV

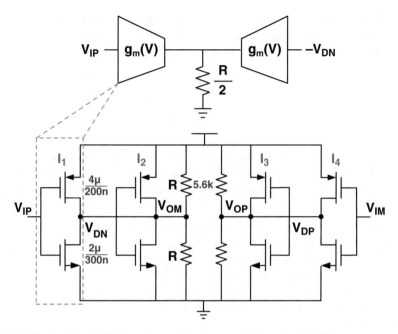

Fig. 7.5 Circuit schematic for the preamp and subtractor within each sub-ADC

1. **step 1** : compare $(V_{IP} - V_{IM}) A_1 - (V_{DP} - V_{DM}) A_2$
   $-0.8 \times A_1 - 0 \times A_2 < 0$
   Although the signal input transistors are in triode and they have lower current gain A_1 compared to the current gain (A_2) of I_2 and I_3, the comparator decision is going to be right in this conversion step as we compare the input with zero.
2. **step 2** compare $(V_{IP} - V_{IM}) A'_1 - (V_{DP} - V_{DM}) A'_2$
   $-0.8 \times A'_1 + 0.5 \times A'_2 < 0$
   Since the drain voltages change the current gain of the transistors change in inverters as well. The comparator decision is going to be correct only if

$$\frac{A'_1}{A'_2} > \frac{0.5}{0.8} \tag{7.1}$$

$$\frac{A'_1}{A'_2} > 0.5 \qquad \text{for full rail to rail input} \tag{7.2}$$

   This is easily true as the input inverter (I_1 and I_4) are in triode and the DAC inverters (I_2 and I_3) are in saturation.
3. **step 3** compare $(V_{IP} - V_{IM}) A'_1 - (V_{DP} - V_{DM}) A'_2$
   $-0.8 \times A'_1 + 0.75 \times A'_2 < 0$

Since the drain voltages change the current gain of the transistors change in inverters. The comparator decision is going to be correct only if

$$\frac{A_1'}{A_2'} > \frac{0.75}{0.8} \tag{7.3}$$

$$\frac{A_1'}{A_2'} > 0.75 \qquad \text{for full rail to rail input} \tag{7.4}$$

At this point the output voltage is near 0.5 V and hence all the transistors (from simulation) are in saturation. However since the gate source voltage of input is smaller in comparison to the DAC, the inverter currents I_1, I_4 are larger than I_2 and I_3.

This steps further continue till we resolve all the bits. If the current increases with the increasing voltage, the comparator will result in a correct decision. Figure 7.8 shows the plot of the current from the signal input transconductance and DAC input transconductance. Initially the differential DAC voltage is zero, leading to the current difference of approximately 500 μA. During the conversion steps, the current difference reduces and finally becomes zero only when the signal input equals the DAC input.

The mismatch between the resistors (R) will result in the gain error between the differential path resulting in common mode components. The comparator after the preamp rejects the common mode component resulting in only affecting the gain of the preamp slightly. Monte Carlo simulation is performed on the resistors and the matching (standard deviation) between the resistors is found to be 0.2 %.

7.4.1   Input Voltage Range

Although the input transistors of the inverters in the g_m cells see rail-to-rail swing, it remains in the saturation region for the entire input range. The input voltage is connected to the inputs of I_1 and I_4 and the inverted DAC output is connected to the inputs of I_2 and I_3. At the end of a SAR conversion the DAC voltage will be close to the inverse of the sampled input value. For example if we sample an input on I_1 of 1 V, the DAC voltage at the input of I_2 will be 0 V near the end of conversion, maintaining an output common mode voltage at nearly a constant 0.5 V. Since the threshold voltage of the transistors is around 450 mV, the transistors in the preamp will not enter the linear region for a peak to peak swing of 950 mV. This is evident from the simulated preamp gain for the whole input swing in Fig. 7.6. The gain of the preamp remains greater than 1 for input swings up to 1.9 Vpp differential.

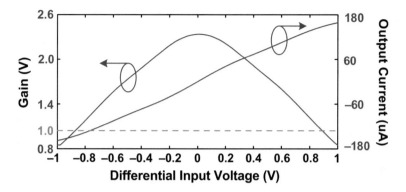

Fig. 7.6 Simulated preamp gain and transconductance output current

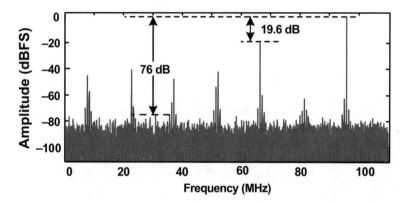

Fig. 7.7 Simulation of the impact of g_m cell non-linearity upon the ADC (*blue*), and the effect when the DAC is predistorted by the same g_m cell (*red*)

7.4.2 Preamp Transconductance Linearity

It is critical that any nonlinearity of the preamp should not alter the digital value. To accomplish this only two things are required: monotonicity and matching the distortion in the input and DAC feedback paths. Figure 7.6 shows the simulated preamp gain and the g_m cell output current versus differential input voltage. As can be seen the g_m cells are clearly non-linear. Since the signal swings rail-to-rail at the input of the preamp, the gain of the preamp changes with the input. The preamp gain increases as the signal amplitude reduces and reaches a peak value of 2.4 for small input differences. Preamp gain is less than one for signal amplitudes greater than ± 950 mV and is symmetric about zero. The simulated output current is shown in red in Fig. 7.6. As can be seen on the RHS of Fig. 7.6, the output current is monotonic with the differential input voltage. The red and blue spectrums in Fig. 7.7 show the simulated linearity of the ADC output with and without DAC

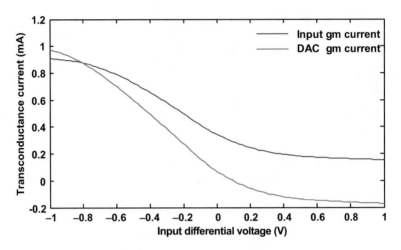

Fig. 7.8 Transconductance current of the signal input and the DAC input

predistortion. The open loop linearity of the inverter transconductance is 19.6 dB over the full rail-to-rail input swing and it increases to 76 dB when the DAC is predistorted by the same inverter transconductor. That is to say distorting both paths by the same nonlinearity in the comparison results in minimal distortion. Unlike traditional SARs, in this design the outputs of the g_m cell (Fig. 7.8) are compared rather than the input and the DAC voltages. If the transfer function of the g_m cell is written as $f(V)$, then as seen in Eq. (7.5), the comparison still holds if both the input and the DAC passes through identical g_m nonlinearities since $f^{-1}f(V) = V$. This is further verified by simulations (Fig. 7.7). The blue spectrum trace with strong nonlinearity results when only the input is passed through the g_m cell and the red trace shows what happens when both input and DAC values are passed through similar g_m cells. The improvement in the linearity from blue to red is attributed to the non-linearity cancellation (effectively DAC predistortion) in the preamp as discussed in Eq. (7.5).

$$V_{IN} < V_{DAC} \equiv f(V_{IN}) < f(V_{DAC}) \equiv V_{IN} < f^{-1}(f(V_{DAC})) \qquad (7.5)$$

Clearly, the inverter gain and the linearity are functions of the power supply and temperature. The power supply and/or temperature affects all the four inverters (11–14) by the same amount, thereby cancelling the transconductance non-linearity of the DAC and the input path simultaneously. Simulations were performed on the entire ADC across temperature and power supply and the spectrum at each temperature and power supply in plotted in Figs. 7.11 and 7.12. The effects of device mismatch were studied with 50 Monte Carlo simulations of the ADC and is plotted in Fig. 7.13. The mean SNDR is 50.11 dB with a standard deviation is 2.11 dB.

7.4.3 Input Capacitance Linearity

Due to the small input sampling capacitance, the input capacitance of the preamp and the parasitics of the sampler becomes significant (12.7 %). Although a constant-V_{gs} sampler reduces the switch nonlinearity, it demands a wider switch, which in turn increases the nonlinear capacitance and the impact of subthreshold leakage. The input and DAC feedback paths have been separated in our design with the result that any capacitance nonlinearity in the input path has no impact on the overall nonlinearity provided the input settles completely. This not true for the DAC feedback path.

Figure 7.10 shows the capacitance variation of NMOS, PMOS and inverter-based g_m cells as a function of input voltage. The transconductance of the PMOS and NMOS transistors are equal in the inverter-based g_m cell. A signal dependent variation in the switch resistance or a variation in the sampling capacitor gives rise to sampler non-linearity in the case of incomplete settling. In this design the switch resistance varies from 134 to 166 Ω and sampling capacitor varies from 148.5 to 154.2 fF over the entire input range. The switch is sized such that the worst case settling error is less than 0.5 LSB. By appropriate sizing of the PMOS and NMOS devices in the inverter the total variation in the capacitance can be reduced to 2 %. On the other hand, the variation in an NMOS-only or a PMOS-only based g_m cell (as might be seen in a differential pair) is 9.9 and 6.9 % respectively. There is a 3.5–5X reduction in the input capacitance variation due to the inverter based subtractor used here.

The nonlinear input capacitance of the preamp scales the capacitive DAC output voltage resulting in static non-linearity (INL) in the ADC. A larger DAC capacitance reduces the impact of the nonlinear input capacitance. Hence the total DAC capacitance is sized based on the ADC INL requirements. It should be noted that increasing the DAC array capacitance does not require an increase in the sampling capacitor as they come into two independent g_m cells as shown in Fig. 7.4.

7.4.4 Gate Leakage

Gate leakage becomes critical for small sampling capacitors in GP technologies as the oxide is thinner. In our design the size of the inverters were increased to reduce the mismatch between them. To validate this the simulated droop in the sampled voltage due to gate leakage is shown in Fig. 7.9. The leakage current causes a change in the input voltage during the conversion time with the maximum droop occurring at the end of the sampling period. Hence the droop in the sampled voltage is simulated at the end of the sampling period, i.e., 9 ns. The maximum droop is 2.8 mV (1.5 LSBs) and only occurs when the input voltage is ±0.95 V. For inputs smaller than ±0.75 V the droop is less than 0.5 LSB. Further gate leakage is a function of the input voltage and hence it results in nonlinearity. The size of the

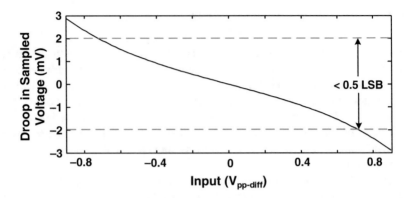

Fig. 7.9 Simulated maximum droop in the sampled voltage vs. the input voltage ($V_{pp\text{-}diff}$) due to gate leakage at 27 °C

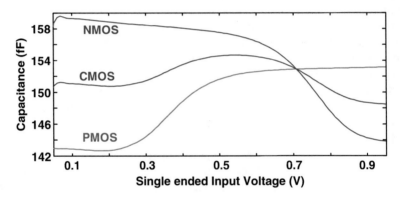

Fig. 7.10 Variation of capacitance with input voltage for NMOS (*blue*), PMOS (*green*) and CMOS implementation for equal transconductance (*red*)

preamp inverters is a compromise between device mismatch, gate leakage and input capacitance nonlinearity. Offset between the inverters results in nonlinearity because it results in incomplete cancellation of the nonlinearity between the input and DAC inverters in the preamp. The preamp inverters are layed-out with one dimensional common centroid geometry with dummies to limit systematic mismatch between transistors.

7.5 Measurement Results

The proposed SAR ADC whose micrograph is shown in Fig. 7.21 was fabricated in TSMC's 65 nm GP process, occupies an area of 0.0338 mm^2 and was packaged in a 5 × 5 mm QFN package.

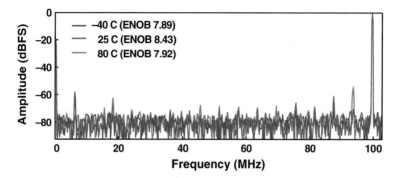

Fig. 7.11 Simulated output spectrum of the ADC at different temperatures

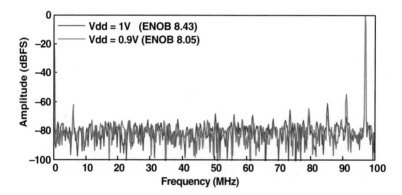

Fig. 7.12 Simulated output spectrum of ADC with 10 % variation in power supply

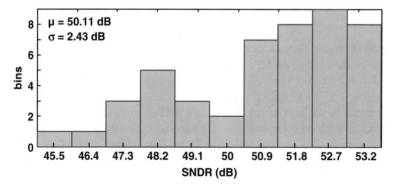

Fig. 7.13 Monte Carlo simulation on the ADC over 50 trials

Figure 7.19 show the measurement setup of the time interleaved ADC. A single ended signal generator is interfaced using $50\,\Omega$ matched baluns. A sixth order passive filter (Fig. 7.20) is used to clean up the harmonics from the signal generator. The bondwires are estimated to have inductance of $1\,nH$. The input is terminated

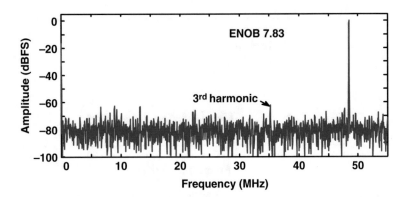

Fig. 7.14 Measured output FFT spectrum of sub-ADC

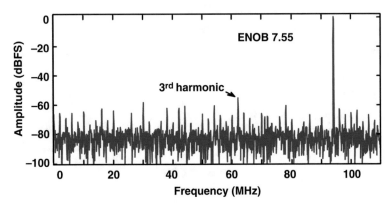

Fig. 7.15 Measured output FFT spectrum of TI-ADC

using 50 Ω terminations on chip. Use of 50 Ω termination on chip rather than on PCB deQs the bondwire thereby reducing bondwire ringing. Figure 7.14 shows the sub-ADC output spectrum for a 49.17 MHz input sampled at 110 MHz. The ENOB obtained is 7.83 bit and the third harmonic is at −60 dBFS. Although we use an active circuit to perform subtraction, we are still able to obtain a linearity of 60 dB due to the nonlinearity cancellation in the preamp g_m cells. Figure 7.15 shows the time-interleaved ADC output spectrum for an input signal of 95 MHz sampled at 220 MHz. The ENOB obtained after gain and offset calibration is 7.55 bit and the third harmonic is at −57 dBFS.

Figure 7.16 shows a plot of the SNDR versus the input amplitude. The peak SNDR is obtained at an input voltage of 1.9 $V_{pp\text{-}diff}$. This gives the maximum amplitude of 95 %. The parasitic capacitance at the DAC output results in an offset which reduces the input dynamic range. Figure 7.17 shows the plot of SNR, SNDR and SFDR with input frequencies. It is seen that SNDR is nearly constant for the

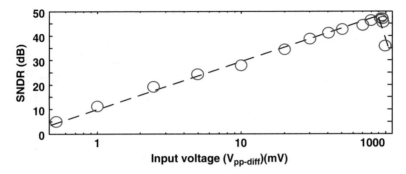

Fig. 7.16 Plot of SNDR vs signal amplitude (V_{pp}) at $f_{in} = 49\,\text{MHz}$

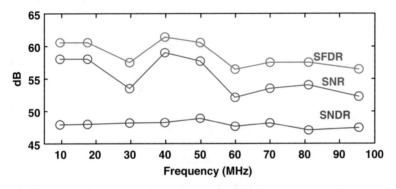

Fig. 7.17 Measured SNR, SNDR and SFDR versus input frequency

entire frequency range as it is limited by the mismatch in the capacitor bank and the inverters in the preamp. The peak SFDR reaches 62 dBFS for a 10 MHz signal and decreases to 57 dBFS at 95 MHz. Figure 7.18a, b show the measured DNL and INL plots for a sub-ADC using a histogram test.

For the input to the ADC, a low frequency sinusoidal signal filtered using passive filters is used. The measured DNL is 1.8/−0.97 LSB and INL is 1.6/−1.5 LSB. Since the DNL without calibration is ±1 bit rather than the desired ±0.5LSB, this is similar to a 6 dB SNDR penalty in the quantization noise. The DNL is greater than ±1 only at the larger input signals. If the differential input swing is greater than 1.9 $V_{pp\text{-}diff}$, the preamp gain is less than one as seen in Fig. 7.6 resulting in higher DNL near maximum code. The nonlinear input capacitance of the preamp causes a nonlinear change in the DAC output voltage. The input parasitic capacitance variation is around 4 fF. A voltage dependent capacitance induces static non-linearity of the DAC characteristic resulting in DAC INL of 1.32 LSB ($4f/770f * 2^8$). This contributes to the measured INL of the ADC (Fig. 7.18). The ADC consumes 844 μW of digital power and 706 μW of analog power resulting in a total of 1.55 mW.

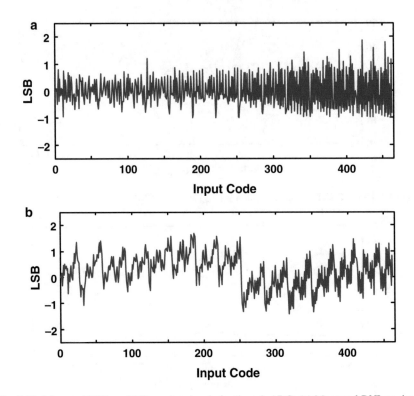

Fig. 7.18 Measured DNL and INL vs. input code for the sub-ADC. (**a**) Measured DNL vs. input code. (**b**) Measured INL vs. input code

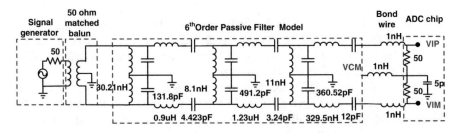

Fig. 7.19 Test bench for testing ADC

Using the figure-of-merit (FOM) definition as shown below, the ADC achieves an FOM of 37.5 fJ/conv-step.

$$FOM = P/(f_s \times 2^{ENOB}) \tag{7.6}$$

Table 7.1 summarizes our design and compares it with other published ADCs with sampling frequencies in this range. While other state-of-the-art ADCs

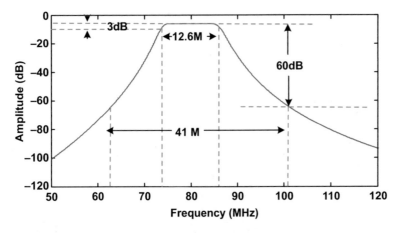

Fig. 7.20 Magnitude response of the simulated passive band pass filter

Fig. 7.21 Micrograph of proposed time-interleaved SAR ADC

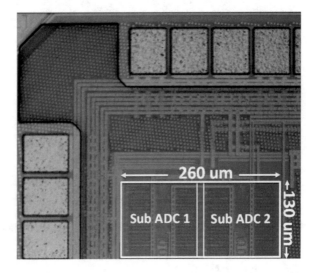

[24, 26, 50, 57] are designed in low power (LP) technologies, we have utilized a general purpose (GP) technology which has more leakage for ease of integration with high performance digital circuits. We get the lowest input capacitance (133 fF) 1.35X lower than [57]. The FOM (37.5 fF/conv) is comparable with the state-of-the-art (Fig. 7.22) and the ADC is more integratable.

Table 7.1 Performance summary and comparison with prior art

	[50]	[57]	[26]	[24]	This work
Tech (nm)	65	65	65	90	65
Area (mm^2)	0.026	0.012	0.044	0.18	0.0338
Supply (V)	1.2	1.2	1.25	1.2	1.0
Input cap (pF)	1.86	0.18	–	2.75	0.13
Fs (MS/s)	100	100	80	100	220
Resolution (bit)	10	9	12	10	9
ENOB (bit)[a]	9.04	8.53	9.4	8.6	7.55
Power (mW)	1.13	1.46	3.46	3	1.55
FOM (fJ/conv)[a]	21.5	50	111.3	55	37.5

[a]ENOB and FOM are calculated at Nyquist frequency

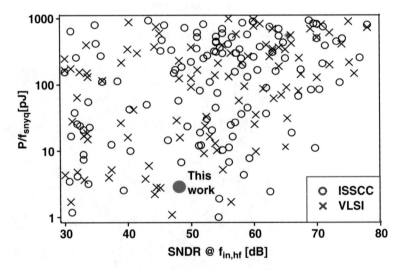

Fig. 7.22 Comparison of prototype ADC with state of art ADCs over energy

7.6 Conclusion

This ADC uses active current subtraction and achieves over 57 dB of linearity by matching input and DAC feedback path nonlinearity. The ADC consumes 1.55 mW power and occupies 0.0338 mm^2 in a 65 nm process. The measured performance corresponds to an FOM of 37.5 fJ/conv-step, which is comparable with the best published ADCs while operating in a GP process. The proposed architecture is significantly more system-compatible owing to its low input capacitance and minimized gate leakage. These two attributes make this design particularly suitable for applications that require a large level of interleaving and where a large number of

ADCs and accompanying drivers are used. Separating the sampling and DAC array capacitors allows us to reduce the sampling time which effectively increases the maximum speed of fully synchronous time-interleaved SAR ADC that is possible. For example, a fully flushed out design with 10X interleaving of our prototype would have resulted in a 9-bit 1.1 GS/s ADC.

References

1. Andrea Baschirotto and Robert Neff. Analog scaling. *isscc 2006*.
2. Behzad Razavi. Design of Analog CMOS Integrated Circuits. *McGraw Hill*, 2000.
3. S.Chatterjee, Yannis Tsividis, and Peter Kinget. A 0.5v analog circuit techniques and their application in OTA and Filter Design. *IEEE Journal of Solid State Circuits*, pages 2373–2387, December 2005.
4. Troy Stockstad and Hirokazu Yoshizawa. A 0.9V 0.5uA Rail-to-Rail CMOS Operational Amplifier. *IEEE Journal of Solid State Circuits*, pages 286–292, March 2002.
5. Keishi Komoriyama, Eiichi Yoshida, Makoto Yashiki, and Hiroshi Tanimoto. A very wideband fully balanced active RC polyphase filter based on CMOS inverters in 0.18 um CMOS technology. *IEEE Symposium on VLSI circuits*, pages 98–99, 2007.
6. Bram Nauta. A CMOS Transconductance-C Filter Technique for Very High Frequencies. *IEEE Journal of Solid State Circuits*, pages 142–153, February 1992.
7. Artur Lewinski and Jose Silva Martinez. OTA Linearity enhancement technique for high frequency applications with IM3 below -65 db. *IEEE Custom Integrated Circuit Conference*, pages 9–12, September 2003.
8. Tien-Yu Lo and Chung-Chih Hung. 1.5-V linear CMOS OTA with −60db IM3 for high frequency applications. *IEEE Asian Solid state Circuit conference*, pages 167–170, 2006.
9. Bahmani F and Sanchez-Sinencio E. A highly linear pseudo-differential transconductance. *IEEE European Solid-State Circuits Conference*, pages 111–114, 2004.
10. A.N Mohieldin, E. Sanchez-Sinencio, and Jose Silva Martinez. Nonlinear effects in pseudo differential otas with cmfb. *IEEE Transactions on Circuits and Systems II*, pages 762–770, October 2003.
11. Benjamin Hershberg, Skyler Weaver, Kazuki Sobue, Seiji Takeuchi, Koichi Hamashita, and Un-Ku Moon. Ring Amplifiers for Switched-Capacitor Circuits. *IEEE International Solid-State Circuits Conference*, 2012.
12. Pietro Andreani and Sven Mattisson. On the use of Nauta's Transconductor in Low-Frequency CMOS g_m-C Bandpass Filters. *IEEE Journal of Solid State Circuits*, pages 114–124, February 2002.
13. S. Vlassis. 0.5 V CMOS inverter-based tunable transconductor. *Analog Integrated Circuits and Signal processing*, pages 289–292, 2012.
14. Barthelemy H, Meillere S, Gaubert J, Dahaese N, and Bourdel S. OTA based on CMOS inverters and application in the design of tunable bandpass filter. *Analog Integrated Circuits and Signal processing*, pages 169–178, 2008.

© Springer International Publishing AG 2017

R.K. Palani, R. Harjani, *Inverter-Based Circuit Design Techniques for Low Supply Voltages*, Analog Circuits and Signal Processing,
DOI 10.1007/978-3-319-46628-6

15. Barthelemy H, Meillere S, Gaubert J, and Kussener E. Transconductance CMOS inverter based AC coupling amplifier. *IEEE New Circuits and Systems Conference*, 2014.
16. Witold Machowski and Jacek Jasielski. Offset Compensation for Voltage and Current Amplifiers with CMOS Inverters. *Mixed Design of Integrated Circuits and Systems (MIXDES)*, pages 382–385, 2012.
17. Rakesh Kumar Palani, Martin Sturm, and Ramesh Harjani. A 1.56 mw 50 MHz 3rd-Order Filter with Current-Mode Active-RC Biquad and 33 dbm IIP3 in 65 nm CMOS. *IEEE Asian Solid state Circuit conference*, November 2013.
18. Rakesh Kumar Palani and Ramesh Harjani. High Linearity PVT Tolerant 100MS/S Rail-to-Rail ADC Driver with Built-in Sampler in 65 nm CMOS. *IEEE Custom Integrated Circuit Conference*, September 2014.
19. Ramesh Harjani and Rakesh Kumar Palani. Design of PVT Tolerant Inverter Based Circuits for Low Supply Voltages. *IEEE Custom Integrated Circuit Conference*, September 2015.
20. Shanthi Pavan. A Fixed Transconductance bias technique for CMOS Analog Integrated Circuits. *IEEE International Symposium on Circuits and Systems*, 2004.
21. Shanthi Pavan and Prabu Sankar. Power Reduction in Continuous-Time Delta-Sigma Modulators Using the Assisted Opamp Techniques. *IEEE Journal of Solid State Circuits*, pages 1365–1379, July 2010.
22. Tamba M, Shimizu A, Munakata H, and Komuro T. A method to improve SFDR with random interleaved sampling method. *International Test Conference*, pages 512–520, November 2001.
23. B Robert Gregoire and Un-Ku Moon. An Over 60-dB True Rail-to-Rail Performance Using Correlated Level Shifting and an opamp with 30 dB Loop Gain. *IEEE International Solid-State Circuits Conference*, 2008.
24. Yan Zhu, U-Fat Chio, Sai-Weng Sin, Seng-Pan U, Rui Paulo Martins, and Franco Maloberti. A 10 bit 100MS/s reference-free SAR ADC in 90 nm CMOS. *IEEE Journal of Solid State Circuits*, pages 1111–1121, June 2010.
25. Brian Drost, M. Talegaonkar, and P. K. Hanumolu. A 0.55V 61 dB-SNR 67 dB SFDR 7 MHz 4th order Butterworth filter using ring-oscillator based integrators in 90 nm CMOS. *IEEE International Solid State Circuit Conference*, pages 360–362, February 2012.
26. Hideo Nakane, Ryuichi Ujiie, Takashi Oshima, Takaya Yamamoto, Keisuke Kimura, Yuichi Okuda, Kosuke Tsuiji, and Tatsuji Matsuura. A fully integrated SAR ADC using digital correction technique for triple-mode mobile transceiver. *IEEE Asian Solid State Circuit Conference*, pages 73–76, November 2013.
27. Michael Figueiredo, Rui Santos-Tavares, Edinei Santin, Joao Ferreira, Guiomar Evans, and Joao Goes. A two-stage Fully Differential Inverter-based self-biased CMOS amplifier with high efficiency. *IEEE Transactions on Circuits and Systems-I*, pages 1591–1603, July 2011.
28. Siva V. Thyagarajan, Shanthi Pavan, and Prabu Sankar. Low distortion active filters using the gm-assisted OTA-RC technique. *Proc. ESSCIRC*, pages 162–165, September 2010.
29. Y. Tsividis. Integrated continuous-time filter design - an overview. *IEEE Journal of Solid State Circuits*, pages 166–176, March 1994.
30. Tien-Yu Lo and Chung-Chih Hung. A 1-V Gm-C low-pass filter for UWB wireless application. *IEEE Asian Solid State Circuit Conference*, pages 277–280, November 2008.
31. Stefano D'Amico, Matteo Conta, and Andrea Baschirotto. A 4.1 mW 10 MHz fourth-order source-follower based continuous-time filter with 79 dB DR. *IEEE Journal of Solid State Circuits*, pages 2713–2719, December 2006.
32. Le Ye, Congyin Shi, Huailin Liao, Ru Huang, and Yangyuan Wang. Highly power-efficient active RC filters with wide bandwidth-range using low-gain push-pull opamps. *IEEE Transactions on Circuits and Systems-I*, pages 95–107, January 2013.
33. S. Kousai, M. Hamada, R. Ito, and T. Itakura. A 19.7MHz, fifth-order active-RC Chebyshev LPF for draft IEEE802.11n with automatic quality-factor tuning scheme. *IEEE Journal of Solid State Circuits*, pages 2326–2337, November 2007.
34. Tore Ulversoy. Software defined radio: Challenges and opportunities. *IEEE Communication surveys and tutorials*, pages 531–550, April 2010.

35. Asad A Abidi. The Path to the Software-Defined Radio Receiver. *IEEE Journal of Solid State Circuits*, pages 954–966, May 2007.
36. Ranim Bagheri, Ahmad Mirzaei, Mohammad E. Heidari, WiLinx Saeed Chehrazi, Minjae Lee, Mohyee Mikhemar, Wai K Tang, and Asad A Abidi. Software-Defined Radio Receiver:Dream to Reality. *IEEE Communications Magazine*, pages 111–118, August 2006.
37. Tonse Laxminidhi, Venkata Prasadu, and Shanthi Pavan. Widely Programmable High-Frequency Active RC Filters in CMOS Technology. *IEEE Journal of Solid State Circuits*, pages 327–336, February 2009.
38. Hesam Amir-Aslanzadeh, Erik J Pankratz, and Edgar Sanchez-Sinencio. A 1-V +31 dbm IIP3, Reconfigurable, Continuously Tunable, Power-Adjustable Active-RC LPF. *IEEE Journal of Solid State Circuits*, pages 495–508, February 2009.
39. Athanasios Vasilopoulos, Georgios Vitzilaios, Gerasimos Theodoratos, and Yannis Papananos. A Low-Power Wideband Reconfigurable Integrated Active-RC Filter With 73 dB SFDR. *IEEE Journal of Solid State Circuits*, pages 1997–2008, September 2006.
40. Stefano D'Amico, Vito Giannini, and Andrea Baschirotto. A 4th-Order Active-Gm-RC Reconfigurable (UMTS/WLAN) Filter. *IEEE Journal of Solid State Circuits*, pages 1630–1637, July 2006.
41. Vito Giannini, Jan Craninckx, Stefano D'Amico, and Andrea Baschirotto. Flexible Baseband Analog Circuits for Software-Defined Radio Front-Ends. *IEEE Journal of Solid State Circuits*, pages 1501–1512, July 2007.
42. Kong-Pang Pun, Chiu-Sing Choy, Cheong-Fat Chan, and J.E da Franca. Digital frequency tuning technique based on current division for integrated active RC filters. *Electronic Letters*, 2003.
43. Hyunchol Shin and Youngcho Kim. A CMOS Active-RC Low-Pass Filter With Simultaneously Tunable High- and Low-Cutoff Frequencies for IEEE 802.22 Applications. *IEEE Transactions on circuits and systems II*, pages 85–89, 2010.
44. Rakesh Kumar Palani and Ramesh Harjani. A 4.6 mW, 22 dBm IIP3 all MOSCAP Based 34–314 MHz Tunable Continuous Time Filter in 65 nm. *IEEE Custom Integrated Circuit Conference*, September 2015.
45. Tien-Yu Lo and. A wide tuning range Gm-C filter for multi mode direct-conversion wireless receivers. *IEEE European Solid State Circuit Conference*, pages 210–213, September 2007.
46. J N Kuppambatti et.al. A 0.6 v 70 MHz 4th-Order Continuous-Time Butterworth Filter with 55.8 dB SNR, 60 dB THD at +2.8 dbm Output Signal Power. *IEEE International Solid-State Circuits Conference*, pages 302–303, 2014.
47. et al M.S Oskooei. A CMOS 4.35-mW +22-dBm IIP3 Continuously Tunable Channel Select Filter for WLAN/WiMAX Receivers. *IEEE Journal of Solid State Circuits*, pages 1382–1391, September 2009.
48. et al Jeffrey Harrison. A 500MHz CMOS Anti-Alias Filter using Feed-Forward Op-amps with Local Common-Mode Feedback. *IEEE International Solid-State Circuits Conference*, 2003.
49. Walt kester. Which ADC Architecture Is Right for Your Application. http://www.analog.com/analogdialogue.
50. Chun-Cheng Liu, Soon-Jyh Chang, Guan-Ying Huang, Ying-Zu Lin, Chung-Ming Huang, Chih-Hao Huang, Linkai Bu, and Chih-Chung Tsai. "A 10b 100 MS/s 1.13 mW SAR ADC with Binary-Scaled Error Compensation". *IEEE International Solid State Circuit Conference*, pages 386–387, 2010.
51. Harpe PJ A, Zhou C, Yu Bi, vander Meijs NP, Xiaoyan Wang, Philips K, Dolmans G, and de Groot H. A 26 μ W 8 bit 10MS/s Asynchronous SAR ADC for Low Energy Radios. *IEEE Journal of Solid-State Circuits*, pages 1585–1595, Jul 2011.
52. Akira Shikata, Ryota Sekimoto, Tadahiro Kuroda, and Hiroki Ishikuro. A 0.5V 1.1MS/sec 6.3fj/conversion-step SAR ADC with Tri-Level Comparator in 40 nm CMOS. *IEEE Symposium on VLSI circuits*, pages 262–263, 2011.

53. Yasuhide Kuramochi, Akira Matsuzawa, and Masayuki Kawabatta. A $0.05\,\text{mm}^2$ $110\,\mu\text{w}$ 10-b self calibrating successive approximation adc core in $0.18\,\mu\text{m}$ cmos. *IEEE Asian Solid State Circuits Conference*, pages 224–227, 2007.
54. Yanfei Chen, Xiaolei Zhu, Hirotaka Tamura, Masaya Kibune, Yasumoto Tomita, Takayuki Hamada, Masato Yoshioka, Kiyoshi Ishikawa, Takeshi Takayama, Junji Ogawa, Sanroku Tsukamoto, and Tadahiro Kuroda. Split Capacitor DAC Mismatch Calibration in Successive Approximation ADC. *IEEE Custom Integrated Circuits Conference*, pages 279–282, 2009.
55. Shuo-Wei Michael Chen and Robert W Brodersen. A 6-bit 600 MS/s 5.3 mW Asynchronous ADC in 0.13 um CMOS. *IEEE Journal of Solid State Circuits*, pages 2669–2680, December 2006.
56. M Hesener, T. Eicher, A Hanneberg, D. Herbison, F Kuttner, and H Wenske. A 14b 40MS/s Redundant SAR ADC with 480 MHz Clock in 0.13 um CMOS. *IEEE International Solid State Circuit Conference*, pages 248–249, February 2007.
57. Yanfei Chen, Sanroku Tsukamato, and Tadahiro Kuroda. "A 9b 100 Ms/s 1.46 mW SAR ADC in 65 nm CMOS". *IEEE Asian Solid State Circuit Conference*, pages 145–148, 2009.
58. R. Vitek, E. Gordon, S. Maerkovivh, and A. Beidas. A $0.015\,\text{mm}^2$ 63 fJ/conversion-step 10 bit 220 MS/s SAR ADC with 1.5b/step redundancy and digital metastability correction. *IEEE Custom Integrated Circuit Conference*, 2012.
59. Y. Zhu, C.H. Chan, S.W. Sin, S.P. U, and R.P. Martins. A 35fJ 10b 160MS/s Pipelined-SAR ADC with decoupled flip-around MDAC and self-embedded offset cancellation. *IEEE Asian Solid State Circuit Conference*, pages 61–64, 2011.
60. Rakesh Kumar Palani and Ramesh Harjani. A 220MS/s 9bit 2X Time-Interleaved SAR ADC with a 133fF Input Capacitance and a FOM of 37fJ/conv in 65 nm CMOS. *IEEE Transactions on Circuits and Systems-II*, 2015.
61. Doris Schmitt-Landsiedel Jens Sauerbrey and Roland Thewes. A 0.5-V 1-uW Successive Approximation ADC. *IEEE Journal of Solid-State Circuits*, pages 1261–1265, July 2003.
62. Peter Kinget. Device mismatch and tradeoffs in the design of analog circuits. *IEEE Journal of Solid-State Circuits*, pages 1212–1224, June 2005.

Printed in the United States
By Bookmasters